Nothing to Fear

A Bright Future with Fossil Fuels

Dears, Donn D.

Nothing to Fear / A Bright Future with Fossil Fuels

Includes index

1. Greenhouse Gasses
2. CO2 Emissions
3. Cap & Trade
4. Coal-Fired Power
5. Renewable Energy
6. Fossil Fuels

ISBN 978-0-9815119-2-4

Published by the Critical Thinking Press

Manufactured in the United States

© 2015 by Donn D. Dears LLC

Dedicated to:

Marion, the love of my life who is missed beyond belief.

Table of Contents

Figures   Page

"The only thing we have to fear is fear itself."

Franklin Delano Roosevelt
First Inaugural Address, 1933

Foreword

Nothing to Fear

Nothing to Fear explains why mankind has the ability to withstand nearly everything mother nature may throw at it, so long as mankind doesn't institute policies that cripple its ability to respond to potential threats.

Mankind needs the availability of more energy that's both affordable and reliable, if it is to have the ability to respond to severe weather, heat or cold, rising seas, drought or any of the other conditions that could afflict people worldwide.

The terrible conditions that plagued mankind in the past, when it lacked the energy to protect itself, included famines, starvation, death from freezing cold, and death from diseases that are now largely under control.

Recently, policies have been implemented that undermine mankind's ability to withstand threats Mother Nature may thrust upon it.

These policies have been instituted as the result of an unreasonable fear of carbon dioxide, an invisible harmless gas released when fossil fuels are burned, and an unjustified terror of climate change.

Unreasonable fear and unjustified terror were phrases used by president Roosevelt in his inaugural address to highlight that fear itself is a threat.

Today, we face a threat conjured up from questionable science that uses fear to promote policies that undermine mankind's ability to react to events.

1

Foreword

These policies are failing, and more importantly, crippling our ability to fend off any problem that might result from changes in our natural world, whether it be higher temperatures, colder temperatures, drought or sea level rise.

The fact is, fossil fuels provide the affordable and reliable energy needed to protect mankind from Mother Nature's whims.

Nothing to Fear addresses each of the policies adopted as a result of the unjustified fear of carbon dioxide. Facts are used to establish why these policies are failing, with specific examples, such as from California.

Nothing to Fear uses data presented by the California Independent System Operator to illustrate the effect renewables have on the ability of the California grid to absorb renewable energy. The graph, which resembles a duck, explains why wind and solar will fail.

But that's only part of the story: Why are we fiddling with renewables in the first place?

Why are we instituting policies that hurt mankind's ability to protect itself from Mother Nature?

The answer to this question begins earlier, with people like Paul Ehrlich and Bill McKibben.

Ehrlich, in *The Population Bomb,* lamented that there were too many people.

A reviewer of McKibben's book, *The End of Nature,* summarized McKibben's message this way: "Until such time as homo sapiens decide to rejoin nature, some of us can only hope for the right virus to come along."

Foreword

But Alex Epstein, in *The Moral Case for Fossil Fuels,* exposes the anti-human, anti-mankind thesis of the radical Greens: The primary criteria for evaluating energy should be whether its production and use benefits or harms mankind.

The book starts with a brief history of the global drive for cutting CO2 emissions, with Chapters 1, 2, and 3 describing the Earth Summit in Rio De Janeiro in 1992, creation of the IPCC in 1988 and passage of legislation and regulations in the United States.

Nothing to Fear explores the flawed science behind the social movement that blames humans for global warming and climate change. Discrepancies with the science are discussed in Chapter 4, and the status of CO2 emissions is summarized in Chapter 5.

The problems and limitations of wind and solar energy are presented in Chapters 6,7 and 8.

Chapter 9 discusses what I call the "Duck Curve," shown below. It shows how increased reliance on renewable fuels between now and 2020 will undermine the grid that collects and distributes energy to our homes and businesses.

Chapters 10 and 11 address the false hope of biofuels and how taxpayer-funded subsidies grossly distort energy markets.

Chapters 12, 13 and 14 explain why carbon capture and sequestration lacks credibility, how the Waxman-Markey proposed legislation foretold the implementation of extreme EPA regulations, and why efforts to substantially reduce carbon dioxide emissions are futile.

Foreword

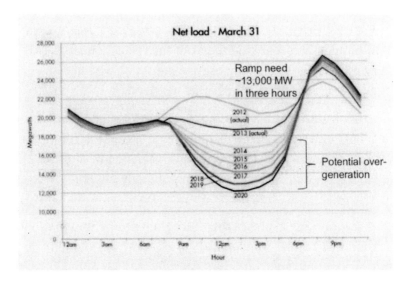

CAISO Duck Curve

To be complete, any book on energy must also discuss possible alternatives to the hypothesis that climate change is caused by carbon dioxide emissions. While there are several such hypotheses, the sun is singled out in Chapter 15, along with the possibility that mankind may be faced with a new Little Ice Age.

Chapters 16 describes the damage that will be done to society, and the harm that will befall millions of people around the world if governments adopt policies that attempt to drastically cut carbon dioxide emissions

Chapter 17 & 18 examine why energy, primarily from fossil fuels, will allow mankind to eliminate the scourges of poverty that remain in dark corners of the globe, and why energy will allow mankind to protect itself from nearly anything mother nature may conjure up in the future.

Foreword

Obviously, little can be done if an asteroid hits the Earth, or if the Yellowstone caldera erupts, but even in these dire situations energy might prove to be mankind's salvation.

My hope is that people will again start to use critical thinking. I believe the failure to develop critical thinking is one of the great failings of our current education system.

The advertising industry relies on superficial thinking for people to accept its message. When an advertisement says, "No other product is better than product A," it's an admission that other products are just as good, products that may be lower in price and a better value than product A.

When an ad claims a product has a value of $200, people should ask, how is the value established? It's not the price tag on the product, but rather what people are actually willing to pay, which may be less than the so-called "special" price being advertised.

I first ran across the poem that has stood me in good stead, in *The Power of Positive Thinking.*

> "I had six honest serving men, they taught me all I knew. They are what and why and when, and where and how and who."

And another that reflects a philosophy of life contrary to the messages put forth by Ehrlich, McKibben and those who are using fear to promote policies harmful to mankind:

> "Two men looked out through the prison bars, one saw the mud, the other the stars."

Foreword

Let's look to the stars and the bright future that lies before mankind as we put the evil message embedded in the carbon dioxide hypothesis of catastrophe behind us.

Part One

The Establishment Takes Aim

"Ninety-seven percent of scientists agree: climate change is real."

President Barack Obama
2015 Tweet

Chapter One

Rio

Temperatures in Rio De Janeiro during June 1992 were a comfortable 80 degrees.

The United Nations Conference on Environment and Development (UNCED) couldn't have found a better location for the conference.

Delegates from 172 governments attended the regular sessions. Many were excited for the opportunity to establish a mechanism for stopping the emission of greenhouse gasses.

There were also 2,400 members of non-governmental organizations (NGOs) that had a unique position at the conference. They were not members of any government delegation, but were given "consultative status" so they could influence the outcome of the meeting.

They supposedly represented the "economic and social life of the peoples of the areas they represented," and were given passes to the conference where they could speak at the conference and at the many committee hearings.

Another 17,000 people attended a parallel meeting that engendered excitement and color for the deliberations taking place at the conference. For the most part, these other attendees can best be described as radical environmentalists with some anti-capitalists and anarchists mixed in.

Each government had one vote, so the interests of the United States were subsumed by the interests of other governments, and, importantly, the influence of the NGOs.

Rio

There were several issues discussed at the conference including toxic substances, biologic diversity, water shortages and global warming.

President Bush was not initially at the conference, but it was an election year, and Senator Al Gore created sufficient political pressure that President Bush was coerced into attending.

The US delegation of 60 was headed by William Reilly, who strongly endorsed a climate convention and strongly supported reducing CO_2 emissions. He had been president of the World Wildlife Fund. In fact, the U.S. delegation was severely divided with shouting matches among the American delegates.

Luminaries, such as Ted Turner and Jane Fonda were there, which added to the drama.

The conference passed the United Nations Framework Convention on Climate Change (UNFCCC), a treaty that put the United States on a path that would have very bad consequences for consumers, workers and taxpayers in the United States.

President Bush signed the UNFCCC treaty on June 12, 1992, it was ratified by the Senate on October 15, 1992, and entered into force on March 21, 1994.

Ratification took place during a heated presidential campaign where most candidates, including President Bush, wanted to have a favorable environmental record.

The UNFCCC also established a Conference of the Parties (COP) where the signatories would hold annual meetings to promulgate new proposals for countering mankind's effect on the planet. As with the Earth Summit in Rio, these meetings are attended by approximately 195 signatory countries under the auspices of the United Nations, plus NGOs.

Since 1994, the United Sates has participated in annual meetings of the Conference of the Parties (COP).

Presumably, the United States Senate ratified the treaty assuming it was a hollow document, where all the actions required by the treaty were voluntary. And indeed, great emphasis was placed on the UNFCCC being entirely voluntary.

This was huge mistake as the United States has been consistently outvoted at the annual COP meetings, and vilified as being recalcitrant, stubborn and an obstructionist, by the other nations who want all nations, especially developed nations, such as the United States, to agree to strict legal commitments for cutting CO_2 emissions and for providing financial assistance to developing countries.

UNFCCC established a bureaucratic structure that could always force the United States into a corner. Each country has a single vote, thus leaving the United States at the mercy of approximately 195 countries, many of whom have agendas hostile to the United States, including Iran, Venezuela and North Korea.

There is no veto, so the United States has virtually no recourse when it is outvoted.

NGOs attend the COP meetings where they play an active, but nonvoting, role. Typically there are more than 10,000 participants at each COP meeting.

An example of how the COP delegates treat the U.S. delegation occurred at the COP 13 meeting in Bali, where the participants booed the U.S. delegation because the delegation didn't want to agree to the Bali Road Map. Fearing negative publicity from the news media the American delegation succumbed to the "consensus."

As a group, these countries have nothing to lose by placing demands on the United States since they can easily outvote the U.S. on every issue.

The COP acts as judge and jury with respect to climate change issues, and has established subsidiary bodies for this purpose, specifically the Subsidiary Body for Implementation (SBI) and the Subsidiary Body for Scientific and Technological Advice (SBSTA).

These groups hold meetings more frequently than COP meetings and NGOs are also allowed to participate in them.

There is a provision for reconciling disputes, but resolution is either through the International Court of Justice or arbitration/ conciliation by a group appointed from among the 195 member parties.

UNFCCC places the United States under the jurisdiction of parties who are inherently unsympathetic toward the United States.

Iran, for example, has signed the treaty and could sit in judgment on the United States.

Parties to UNFCCC should act "on the basis of equity, and in accordance with their differentiated responsibilities and respective capabilities."

Article 4.4 states, "Developed country parties ... shall assist developing country parties that are particularly vulnerable to the adverse effect of climate change in meeting costs of adaptation to those adverse affects."

This theme is repeated throughout UNFCCC. Developed countries (especially the United States) should bear the burden of reducing CO2 emissions and also provide financial support to developing countries.

This concept places the United States in the ridiculous position of contributing money to China, by paying its annual share of the $100 Billion Green Climate Fund (GCF) if it is ever fully funded.

UNFCCC also established a requirement for preparing "educational materials and creating public awareness on climate change and its effects," which amounts to self-promotion and propaganda opposing any dissent to the party line.

The educational material allows UNFCCC to propagandize on behalf of cutting CO2 emissions.

The language of UNFCCC is frequently imprecise, and this has serious implications for the United States.

Under UNFCCC, parties have "the responsibility to ensure that activities within their jurisdiction or control do not cause damage to the environment of other States or of areas beyond the limits of national jurisdiction."

Who is to say whether emissions from American factories are causing damage to the environment of Europe or any other area of the world? At the very least, this statement allows people to pillory the United States because, in their minds, the United States is harming the environment.

Terms such as sustainable development, sustainable economic growth, and sustainable agriculture are not defined in the Convention. Presumably COP meetings, or SBI or SBSTA meetings, will make determinations about these terms and the United States will have little to say about how they will be administered.

UNFCCC was supposed to be voluntary, but has been used to pressure the United States. The Kyoto protocol that emerged from UNFCCC was legally binding and dropped all pretense of volunteerism.

The Kyoto treaty established legally binding emission targets for developed countries, listed in Annex I of UNFCCC. The Kyoto treaty required the United States to reduce its greenhouse gas emissions 7% from 1990 levels.

It was reportedly signed on behalf of the United States by Al Gore, but more likely by a member of the US UN delegation.

Fortunately the Senate, by unanimous vote, had previously passed the Byrd-Hagel resolution saying the United States would not ratify any treaty that didn't require developing countries to also cut CO2 emissions, and as a result, President Clinton never submitted the treat for ratification.

Annex II countries, including the United States if it had ratified the treaty, were obligated to provide funding and insurance to developing countries; including obligations for any perceived impacts on developing countries from actions taken by Annex I countries in meeting their emission targets under the Kyoto protocol.

India and China were not Annex I or II countries so were not required to meet any target for reducing greenhouse gasses and were considered developing countries under UNFCCC and the Kyoto Protocol.

Continuing the bureaucratic nature of the UNFCCC treaty and the Kyoto Protocol, the COP would serve as the Meeting of the Parties, "COP/MOP" for the Kyoto Protocol.

The Bali Road Map established in December 2007, created a process that would culminate in Copenhagen, Denmark where the "fifteenth session of the Conference of the Parties and the fifth session of the COP/MOP would be held from November 30 to December 12, 2009," to arrive at a new treaty to replace the Kyoto Protocol.

The Copenhagen meeting failed and, following several COP meetings, attention is focused on COP 21 in Paris, during December 2015, to establish a treaty to replace the expired Kyoto Treaty. Again, to force mandatory cuts in CO_2 emissions.

The United Nations has demanded that developed countries, including the United States, cut their emissions 80% by 2050.

Having ratified UNFCCC, the United States participation in the COP 21 meeting in Paris is essentially mandatory, again placing the United States under tremendous pressure to agree to whatever emerges.

President Obama has said he strongly favors cutting CO_2 emissions 80% by 2050 and that he wants the United States to assume a leadership role in this regard.

Rio

There was little if any debate in Rio in 1992 about whether humans were causing global warming by burning fossil fuels and otherwise emitting CO2 and other greenhouse gasses.

It was a given ... a Fait Accompli ... because the IPCC, an organization created to give UNFCCC scientific guidance ... said so.

Very few Americans in 1992 had even heard about global warming or climate change, yet the President and Senate agreed to a treaty that would have long lasting, serious consequences for all Americans.

Chapter Two

The IPCC

The International Panel on Climate Change (IPCC) was established by the United Nations, General Assembly Resolution 43/53 of 6 December 1988, which recited the following concerns[1]:

> "Concerned that certain *human activities* could change global climate patterns, threatening present and future generations with potentially severe economic and social consequences ...

> "Noting with concern that the emerging evidence indicates that continued growth in *atmospheric concentrations of "greenhouse" gasses* could produce global warming with an eventual rise in sea levels, the effects of which could be disastrous for mankind if timely steps are not taken at all levels,"

(Emphasis added.)

The IPCC's role was defined by the Principles Governing IPCC Work approved by the General Assembly[2], which stated:

> "The role of the IPCC is to assess on a comprehensive, objective, open and transparent basis the scientific, technical and socio-economic information relevant to understanding the scientific basis of risk of *human-induced climate change*, its potential impacts and options for adaptation and mitigation."

From the very beginning the purpose of the IPCC was NOT to determine the cause of global warming, because, by some magical mechanism, it had already been established that greenhouse gasses and human activity were the cause of global warming and climate change.

This same presumption was evident at the Rio conference and its outcome, the UNFCCC treaty.

Prior to the formation of the IPCC, no government or international organization had set out to determine the cause of global warming.

The hypothesis that man-made CO_2 is responsible for rising global temperatures (or more ambiguously "climate change") originated in the late 1800s and is attributed to Arrhenius, though others had also seen the possibility that greenhouse gasses could cause warming. During the second half of the twentieth century, some individual scientists, the World Meteorology Organization, United Nations Environment Program (UNEP) and non-governmental organizations, explored the possibility that greenhouse gasses could cause global warming, but there was no organized effort to scientifically attribute global warming or climate change to man-made CO_2.[3]

When the IPCC was created, it was merely assumed that atmospheric CO_2 was causing climate change, and alternative possibilities were mostly ignored.

Whether this assumption was merely accidental negligence or deliberate is not unimportant. The individuals prominently associated with the work of the IPCC were widely known to be environmental activists and leftists; they didn't make a secret of their motivation.

The IPCC

The IPCC has issued Assessment Reports beginning with AR1 in 1990. Thus far it has issued five such reports, with the latest in 2014.

A thousand or more scientists contributed to recent reports, though far fewer than that are actual climate scientists or helped write or review the critical chapters on what causes climate to change.

Some scientists have resigned from the IPCC, claiming their views were not incorporated in the final synthesis report.

Most importantly, *government representatives,* not necessarily scientists, approve the all important synthesis report with its summary for policy makers.

While there may have been a thousand scientists who worked on the report, only a small cadre, perhaps 40 people, actually approve the synthesis report and the all important Summaries for Policy Makers.

It is the Summaries for Policy Makers that are read by reporters and others.

Any glaring errors or misrepresentations could be seen by scientists who read the entire report, but reporters and others are not likely to read the complicated reports that contain scientific formulas and jargon.

Here is how one scientist describes how evidence that could negate the CO_2 hypothesis is excluded from the summary for policy makers (SPM).[4]

An example of this SPM deception occurred with the 1995 Report. The 1990 Report and the drafted

1995 Science Report said there was no evidence of a human effect. Benjamin Santer, as lead author of Chapter 8, changed the 1995 SPM for Chapter 8.

Original language:

> "While some of the pattern-base discussed here have claimed detection of a significant climate change, *no study to date has positively attributed all or part of climate change observed to man-made causes.*"

Revised language:

> The body of statistical evidence in chapter 8, when examined in the context of our physical understanding of the climate system, *now points to a discernible human influence on the global climate.*

(Emphasis added.)

"In short, [the SPM] is advocacy, not assessment,"[5]

The graph on the next page from IPCC AR4 shows how CO_2 concentrations in the atmosphere, along with two other greenhouse gasses, have increased beginning with the industrial revolution.

As of today, the atmospheric CO_2 level is approximately 400 ppm.

The purpose of using this particular graph, rather than one that only covers the period from 1800 to today, is to show that CO_2 levels have remained relatively constant at 280 ppm for a long period of time, in this instance back to the birth of Christ.

The IPCC

The United Nations and other sources have claimed that worldwide CO2 emissions must be cut 50% by 2050 if total atmospheric emissions are to be kept below 450 ppm, the point beyond which they say there would be a climate catastrophe.

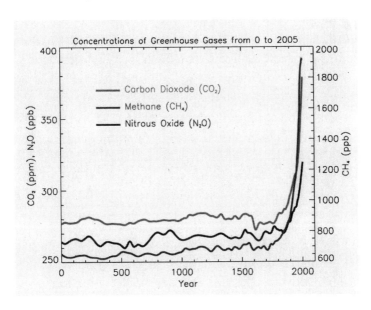

Fig1. Greenhouse gas atmospheric levels from IPCC AR4

To accomplish this objective while allowing developing countries, such as China, to continue to emit CO2, albeit at lower levels, the United States and other developed countries would have to cut their emissions 80% by 2050.[6]

Achieving such massive reductions would require upending the world's economy.

The IPCC

Christiana Figueres, executive secretary of the UNFCCC, made this very clear when she said:

> "This is probably the most difficult task we have ever given ourselves, which is to intentionally *transform the economic development model*, for the first time in human history. This is the first time in the history of mankind that we are setting ourselves the task of intentionally, within a defined period of time, to change the economic development model that has been reigning for the, at least, 150 years, since the industrial revolution." (Emphasis added.)

The world's economic model is based on capitalism.

On the face of it, Figueres made it clear it will be necessary to do away with capitalism to achieve the objective of cutting CO2 emissions.

And Naomi Klein said, "Forget everything you think you know about global warming. It's not about carbon — it's about capitalism."

Earlier in 2010, Ottmar Edenhofer, co-chair of the IPCC Working Group III, said[7]:

> "*Climate policy has almost nothing to do anymore with environmental protection*, the next world climate summit in Cancun is actually an economy summit during which the distribution of the world's resources will be negotiated." Emphasis added.

Clearly the IPCC has an agenda. It's clear the IPCC is more interested in redistributing wealth than actually discovering the causes of climate change.

Chapter Three

Creeping Agreement by the U.S.

The UNFCCC COP 21 meeting in Paris, December 2015, is designed to achieve an agreement whereby:

(1) Developed nations, including the United States, would agree to dramatically cut CO2 emissions

(2) An annual $100 billion development fund would be established, where developed countries, including the United States, would pay into the fund, so the fund administrators could allocate money from the fund to undeveloped nations, including China and India, so they could mitigate the effects of climate change.

The Obama administration worked hard during 2015 to convince Americans that an agreement in Paris at COP 21 to cut CO2 emissions would be good. It used the regulations established by the EPA to demonstrate to other countries that the United States was serious about cutting CO2 emissions.

Many believe that as long as the Senate is controlled by Republicans, programs agreed to in Paris, or at any UNFCCC COP meeting, can't be implemented.

To a certain degree this is true. It is important the Senate not approve any agreement or treaty that comes out of COP 21 in Paris.

Unfortunately, a president set on "transforming" America by destroying the energy infrastructure can act in other ways that don't require an international treaty.

1. First, the president can, and has, issued executive orders, and implemented regulations, primarily through the EPA, requiring the adoption of actions to cut CO_2 emissions.

2. Second, what happens in California doesn't stay in California. There exists a strategy underway today, an infiltration or co-opting strategy, that bypasses Congress. Its intent is to have other states adopt regulations similar to those adopted by California, such as cap-and-trade and renewable portfolio standards mandating the use of wind and solar generated electricity.

There is, therefore, a movement already in place to force Americans to comply with regulations that result in the reduction of CO_2 emissions. It's not necessary for Congress to act.

This point was made clear by Tiffany Roberts, a person who has worked in California's Legislative Analyst's Office, where she authored reports on traditional and alternative energy policies and California's cap-and-trade program, when she spoke at the Tenth International Conference on Climate Change in Washington DC in 2015.

Ms. Roberts pointed out that most people in California are not aware they are subject to a cap-and-trade program that is increasing their costs for electricity and everything that incorporates electricity in the final product or service.

The accompanying picture, and additional comments from Roberts' presentation, shows the pervasiveness of the cap-and-trade legislation and how it even affects companies outside California.

Roberts said there are 650 California companies subject to cap-and-trade, covering 85% of California's economy.

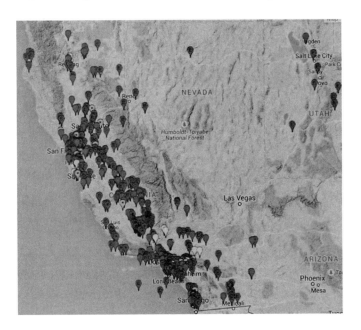

Figure 2, California Businesses Subject to Cap & Trade

California's eleventh cap-and-trade auction brought in around $1 billion to the state's coffers, all of which had to eventually come out of the pockets of ordinary citizens. It's estimated that cap-and-trade revenue will be over $2 billion in 2020, which will also come from the pockets of ordinary citizens.

It's important to recognize that it's people at the lower end of the economic ladder who will be hurt the most, because they use a disproportionate amount of their income for basics, such as gasoline, electricity and food.

A similar cap-and-trade program has been established in New England covering 9 states.

Renewable portfolio standards — laws requiring electric utilities to buy power from solar and wind producers at higher prices than charged by fossil fuel producers of electricity — have been established in 31 states.

The intent is to bypass Congress and go from state to state to implement programs that would meet the CO_2 reduction objectives to be spelled out in any UNFCCC agreement.

This creeping approach to forcing Americans to adopt CO_2 emission cuts without Congressional approval is well underway.

These programs, based on the fear of global warming and climate change, are hurting peopled and causing economic harm.

Chapter Four

Why the CO2 Hypothesis is Wrong

It's not necessary to be a climate scientist to ask questions about the CO2 hypothesis.

It's also not necessary to be a scientist to assess whether the earth has warmed over the past hundred years or so.

The evidence of warming is all around us.

In the early 1900s many rivers and lakes would freeze over, while they no longer do. Some glaciers have receded.

And there are graphs that show global average temperature rise over the past 150 years.

Even so, it's important to separate facts from fiction.

A good example of fiction can be found in Al Gore's movie and book, *An Inconvenient Truth*.

The movie uses Mount Kilimanjaro as the poster child for proof of global warming. It shows that Mt. Kilimanjaro has lost most of its snowcap.

Here's text from the movie, *An Inconvenient Truth*.

> And now we're beginning to see the impact in the real world. This is Mount Kilimanjaro more than 30 years ago, and more recently. And a friend of mine just came back from Kilimanjaro with a picture he took a couple of months ago. Another friend of mine Lonnie Thompson studies glaciers. Here's Lonnie with a sliver of a once mighty glacier.

Why the CO2 Hypothesis is Wrong

Within the decade there will be no more snows of Kilimanjaro.

However there's a problem with this commentary … it isn't accurate if it's meant to show the effect of global warming.

Mount Kilimanjaro is losing its snowcap because of sublimation, not because of warming.[8]

Sublimation is when a solid goes directly to the gaseous state without first becoming a liquid.

Temperatures at the top of Mount Kilimanjaro don't go above freezing so the snow or ice can't melt. As others have shown, there are no rivulets of water flowing from the snowcap.

The probable cause for the loss of ice and snow is a lack of moisture from evaporation due to deforestation of the lower reaches of the mountain.

Another example of irrational thinking was President Obama's use of the Mendenhall Glacier to support his call for action to cut CO2 emissions.

While many glaciers around the world have been retreating due to warming, it does not necessarily mean that CO2 was the cause.

In the case of the Mendenhall Glacier, it started retreating around 1765, more than a decade before the American Revolution, and it retreated more than a mile before the beginning of the Industrial Revolution in the mid-1800s.

For nearly a century after the glacier started its retreat, atmospheric CO2 levels remained at around 280 ppm, not rising above 290 ppm until 1870.

Why the CO2 Hypothesis is Wrong

President Obama was in error when, calling for action to cut CO2 emissions, he said:

> This is as good of a signpost of what we're dealing with when it comes to climate change as just about anything ... It sends a message about the urgency we're going to need to have when it comes to dealing with this.

The Mendenhall Glacier started its retreat at the end of the Little Ice Age, as temperatures began to rise from natural causes.

Atmospheric CO2 levels didn't start their rise for almost a century later.

There are at least three other documented cases where glaciers started their retreat in the mid- to late-1700s, well before the rise in atmospheric CO2 levels.[9]

Nigardsbreen Glacier, Norway
Nisqually Glacier, Mt. Ranier, WA
Sorbreen Glacier, Norway

Before reaching a conclusion that CO2 has created a change in our environment, it's important to be certain some other force isn't at work, such as sublimation and deforestation along the flanks of the mountain in the case of Kilimanjaro's snowcap.

The evidence is clear, the world has gotten warmer over the past hundred years or so.

But this visual evidence of warming does not establish that CO2 is causing the warming.

Too often people will use the visual evidence of warming as proof that CO2 is the cause, yet there is no demonstrated cause and effect.

Why the CO2 Hypothesis is Wrong

We can agree that warming has taken place.

Virtually all scientists agree the world has warmed. But the question is why?

One reason people should be skeptical about the CO2 hypothesis is that it can't stand the test of time. In this case looking back two thousand years, or more.

Figure 1 on page 20 shows this clearly.

The figure shows that atmospheric CO2 levels remained essentially constant at around 280 ppm prior to 1870, going back to the time of Christ.

It's clear that temperatures have risen and fallen over this time period, at least 2,000 years, while CO2 levels have remained essentially constant at 280 ppm.

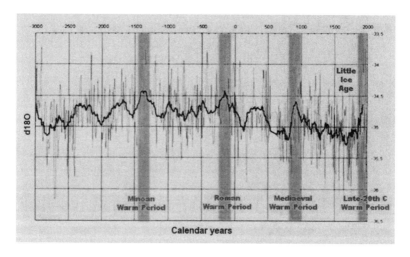

Figure 3. 4,000 Year Temperature History
Temperatures-Solid line
Warm Periods-Shaded

Why the CO2 Hypothesis is Wrong

Figure 3 shows temperature, from Oxygen-18 proxy records, dating back 4,000 years where atmospheric CO2 levels were around 280 ppm for the entire period.[10]

There are many such reconstructions. I chose this one because it has the warm periods clearly marked.

Figure 3 indicates there is little, if any, linkage between atmospheric CO2 levels and temperature.

If there isn't any linkage, the CO2 hypothesis is wrong.

Another way to reach the same conclusion is to look at what has been happening over the past 18 years.

Estimated global temperatures have been essentially constant for the past 18 years, yet atmospheric CO2 levels have continued to rise.

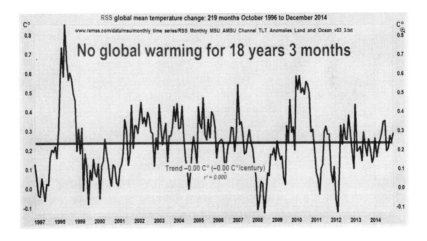

Figure 4. From Christopher Monckton of Brenchley

Why the CO2 Hypothesis is Wrong

The least-squares linear-regression trend on the RSS satellite monthly global mean surface temperature anomaly dataset shows no global warming for 18 years 3 months since October 1996.

Many proponents of the CO2 hypothesis suggest reasons for why temperatures have stopped following CO2 levels. They suggest, for example, that the oceans are absorbing the heat.

No matter what their suggestion, it doesn't refute the fact that global temperature and CO2 levels are not linked.

Figure 3 shows that temperatures are not linked to atmospheric CO2 levels that remained essentially constant at 280 ppm for nearly 4,000 years.

If temperatures during the Mediaeval and Roman Warm Periods were higher, or at least as high as today, while CO2 levels were essentially constant at 280 ppm, there is no linkage between temperatures and atmospheric CO2.

Additional problems with the CO2 hypothesis came from projections made by the IPCC.

These computer projections attempted to forecast future temperatures.

The heavy dotted line in Figure 5 is the average of temperatures derived from all the IPCC computer scenarios.

The bottom two dotted lines are the actual temperatures as determined by two organizations, the University of Alabama in Huntsville, Alabama, and the Climatic Research Unit (University Met Office).

Actual temperatures are below the average forecast by a large margin, and also below all but two of the lowest IPCC scenarios.

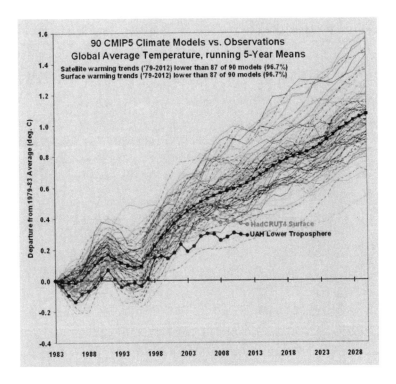

Figure 5. IPCC global temperature forecast compared with actual. From Watts Up With That

Catastrophic climate predictions are based on the IPCC computer projections, not science. And the projections are obviously wrong.

Quoting Vincent Gray, Ph.D., and an IPCC expert reviewer, "No climate model has ever been validated ... Validation must include successful prediction over the entire range of circumstances for which it is required."[11]

Why the CO2 Hypothesis is Wrong

These discrepancies in the IPCC CO2 hypothesis are not trivial.

They demonstrate that the CO2 hypothesis is seriously flawed and that it should not be relied on.

The basic thesis of the CO2 hypothesis is that temperatures rise with an increase in atmospheric CO2 levels.

But temperatures rose during earlier warm periods with no rise in atmospheric CO2 levels. See Figure 3.

This demonstrates that there is no significant linkage between atmospheric CO2 levels and temperatures.

If CO2 and temperatures are not linked, there is nothing to fear and no reason to cut CO2 emissions.

Chapter 5

Current Status of CO2 Emissions

Table 1 shows total global CO2 emissions and the shares contributed by the six largest emitters.

Table 1			
Country	CO2 emissions (MMT)	Per capita emissions (Tons)	CO2 emissions % of Total
World	35,270	-	100.0%
China	10,300	7.4	29.2%
United States	5,300	16.6	15.0%
EU28	3,400	6.8	9.6%
India	2,500	1.9	7.1%
Russia	1,800	12.6	5.1%
Japan	1,400	10.7	4.0%

China now emits the most CO2, roughly twice as much as the United States.

Per capita CO2 is still higher in the United States.

Current Status of CO2 Emissions

The importance of per capita CO2 emissions will be seen when examining the foolhardiness of trying to cut CO2 emissions.

It should be noted that Russia and Japan emit the smallest amounts of CO2, but have the second and third highest per capita emissions.

It's also interesting to note that these six entities emit 70% of the world's CO2.

China and India are both considered developing countries.

China's CO2 emissions in 2007 were about the same as the United States, at around 6,000 Million Metric Tons (MMT).

In the past seven years, China has increased its CO2 emissions by 4,000 MMT. To offset this increase, the United States would have had to cut its emissions by 80%.

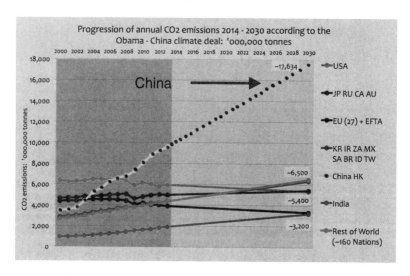

Figure 6. CO2 Emissions 2014 - 2030.
From Watts Up With That by E. Hoskins

Current Status of CO2 Emissions

China has recently said it will limit CO_2 emissions to a peak level to be reached in 2030. At the current rate, as shown by Figure 5, China's emissions in 2030 would be around 18,000 MMT, while the United States would be around 5,400 MMT.[12]

This may be higher than will occur, but an estimate by MIT of 13,000 MMT would seem to be too low.

U.S. CO_2 emissions are likely to be slightly lower than today, assuming the EPA's drastic plans for cutting CO_2 emissions are <u>not</u> put in place.

Lower U.S. CO_2 emissions will probably result from the use of natural gas to replace some coal-fired power plants for economic reasons, assuming natural gas remains below $3.50 per million BTU.

The absurdity of attempting to drastically cut CO_2 emissions will be addressed in Chapter 12.

The distribution of U.S. CO_2 emissions by source are shown in Table 2.

While this data is from 2004, the distribution between sources hasn't changed appreciably. The primary change has been emissions from generating electricity, which have decreased from about 39% to 38%, as the result of replacing coal with natural gas.

Renewables, specifically wind and solar, still only provide a tiny amount of electricity.

Total U.S. CO_2 emissions have declined by about 10% between 2004 and 2014.

Current Status of CO2 Emissions

There are some differences between the data provided by the EPA and the Department of Energy, especially with respect to gasoline and transportation segments.

Table 2

U.S. CO2 Emissions 2004		
Source	MMT	% Total
Electric Generation	2298.6	39%
Gasoline	1162.6	20%
Industrial	1069.3	18%
Transportation (Excluding Gasoline)	771.1	13%
Residential	374.7	6%
Commercial	228.8	4%
Total	5905.1	100%

Total excludes approximately 70 MMT of CO2 emissions from miscellaneous sources.
Source: Emission US Greenhouse Gasses, 2005 by DOE, Energy Information Administration.
MMT = Million Metric Tons

Current Status of CO2 Emissions

Table 2 establishes that emissions from generating electricity and from the use of gasoline provide nearly 60% of total CO2 emissions.

While emissions from industry are also relatively substantial, individual sources are varied. For example, the differences between cement plants, steel mills and manufacturing assembly plants are substantial, which makes it very difficult to establish blanket regulations for cutting industrial emissions.

Blanket regulations can be applied to power plants and automobile exhausts, which is one reason the government has focused on establishing regulations for the electric generation and gasoline segments. For example, restricting CO2 emissions from new coal-fired power plants to 1,400 pounds per megawatt hour and establishing fuel economy standards for automobiles are easily applied to these two segments.

But to establish these regulations the government had to demonstrate there were alternatives for generating electricity and for using gasoline.

Environmental groups had been proposing alternative methods for generating electricity and screaming for an electric car for years.

These meshed with the government's plans for cutting CO2 emissions.

However, these so-called clean energy alternatives cannot produce enough electricity to replace coal and natural gas, and electric cars can't reduce CO2 emissions as long as coal and natural gas are used for generating electricity.

Current Status of CO2 Emissions

In 2014, total U.S. electricity generation was 4,093 billion kWh. It would require essentially doubling U.S. electricity generation to replace gasoline powered cars with battery-powered cars.

The next two chapters explain why these alternatives can't replace coal, natural gas or nuclear for generating electricity.

Chapter 9 explains why the Duck curve demonstrates the negative consequences of attempting to impose the use of renewables, wind and solar.

Part Two

Renewables

"One of the great mistakes is to judge policies and programs by their intentions rather than their results."

Milton Friedman
(1912 - 2006)

Chapter 6

Wind Energy

Total electricity generation in the United States in 2014 was 4,093 billion kilowatt hours.

Of that amount, wind generated 4.4% and solar generated 0.4%.

In total, wind and solar generated 4.8% of the electricity generated in the United States. (This doesn't include PV rooftop solar and some distributed solar from sources such as fuel cells. PV rooftop solar will be covered in the next chapter.)

The increase in the number of wind turbines has been significant over the past dozen years, with approximately 48,000 wind turbines having been installed in the United States.

But this has come at a huge cost to the public that frequently doesn't know it is paying for expensive electricity, or that its taxes have been used to build wind farms.

While it costs the electric utility approximately 5 cents per kWh to generate electricity from natural gas or coal, it costs more than twice this amount to generate electricity from wind.[13] It has also cost approximately $7 billion of taxpayer money to subsidize the construction of these wind farms, not counting local and state subsidies.[14]

These are the visible costs. The hidden costs include the cost of keeping gas turbines in spinning reserve, ready to be brought online whenever the wind stops blowing.

Wind Energy

The hidden costs also include efforts to install storage to obviate the need for backup spinning reserves. At best, storage costs twice what it costs to build a natural gas combined cycle (NGCC) power plant, which is $1,100 / KW.

In addition, there is the cost of building new transmission lines to carry the wind generated electricity from distant areas where the wind is strong, such as along the front range of the Rocky Mountains, to where it's needed in the eastern part of the United States.

There is also the discrepancy between the nameplate rating of wind turbines and the amount of electricity actually produced by the wind turbines. This is established by the capacity factor.

Capacity factor is determined by calculating the amount of electricity the unit actually produces over a year, divided by the amount the unit should theoretically produce based on its nameplate rating.

Over the past few years, the average capacity factor of installed wind farms has been no better than 30%.

While 65,879 MW of wind turbines have been installed through 2014, the amount of electricity they produce is only 30% of what their nameplates say they should produce.

Viewed differently, it would only require around 30 NGCC power plants with a combined rating of 23,000 MW, with each having a capacity factor of 85%, to be able to generate the same amount of electricity as 65,879 wind turbines rated 1 MW each.

In other words, wind farms don't produce very much electricity when compared with NGCC, coal-fired or nuclear power plants. In short, wind turbines are inefficient.

Wind Energy

The main problem with wind energy, however, is that it's unreliable.

Some will say it's unfair to categorize wind as being unreliable. They will say it's intermittent.

But the organizations that manage the grid, and dispatch electricity to the grid, can't rely on wind energy being available when it's needed.

For them, wind is unreliable.

Wind also generates most of its electricity at night, when it isn't needed.

Some wind farm owners have paid the organizations managing the grid to use the electricity generated by the wind farm so the owners could obtain the production tax credit, i.e., subsidy.

A good example of how money is wasted on wind farms is Denmark, one of the countries held up as an example of the use of wind. See Figure 7.

The grey bars at the top represent power from central power stations, which is essentially from fossil fuel power plants.

The horizontal line shows total load.

Clearly, all of the load could have been supplied solely by fossil fuels, possibly with some small assist from combined heat and power (CHP).

The investment in wind turbines has been completely wasted.

Considering that wind is more expensive than fossil fuels in most countries, it makes no economic sense to build wind turbines.

Wind Energy

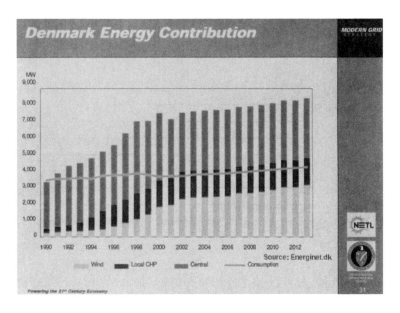

Figure 7. Denmark Wind Energy from NETL

In actuality, Denmark exports a large amount of its wind generated electricity to Norway, because Norway, where over 90% of its electricity is provided by hydro, can absorb the excess wind generated electricity from Denmark, which would otherwise have to be dumped.

Wind also doesn't generate electricity on hot summer days when it's needed for the air conditioning load, because the wind typically doesn't blow on hot, summer days.

The New York Times reported two years ago that the grid almost failed during a heat wave, because the wind farms couldn't provide the needed electricity even though the nameplate ratings indicated they should have been able to.

Wind Energy

From the New York Times:

> Peak supply is also becoming a vexing problem
> because so much of the generating capacity added
> around the country lately is wind power, which is
> almost useless on the hot, still days when air-
> conditioning drives up demand.[15]

Not only are there good reasons for not using wind to generate
electricity, but wind energy can't possibly replace all the electricity
generated by traditional methods.

Today, wind turbines only generate 4.4% of the electricity
produced in the United States.

It would require building another 1,500,000 wind turbines to
generate the same amount of electricity as traditional methods. But
this assumes the wind blows all the time, and it doesn't include
constructing thousands of miles of new transmission lines costing
$100 billion.[16]

Unreliable wind energy drives up the cost of electricity and
deprives Americans and American industry access to the least
costly, most reliable electricity which is generated using fossil
fuels.

Chapter 7

Concentrating Solar Energy

There are basically two approaches for generating electricity from the sun's rays, or photons.

The first is from central facilities using either concentrating solar (CSP) or photovoltaic (PV) solar installations.

The second is using PV rooftop solar installations, which will be discussed in Chapter 8.

The tiny amount of electricity recorded as being produced from solar, or 0.4% of the total electricity produced in the United States in 2014, was from central facilities operated by utilities.

Ivanpah, is an example of a CSP facility located in California.

Figure 8. Photo of Ivanpah from US Department of Energy

Concentrating Solar Energy

Ivanpah uses three solar towers, similar to the one shown in the photograph, with thousands of sun tracking mirrors, heliostats, reflecting the sun's rays, concentrating and focusing them onto a boiler at the top of each tower.

The boiler contains water that becomes steam from the intense heat of the sun's rays. The steam, in turn, drives a turbine generator that generates electricity.

Ivanpah is the largest CSP facility in the United States.

A major problem with CSP is its cost.

Ivanpah cost over $5,600 / KW to build. This is nearly the cost of a new nuclear power plant.

While Ivanpah is viewed as so-called clean energy, it uses natural gas to warm the plant each morning. The State of California has approved Ivanpah to use 525 million cubic feet of natural gas annually.

Another problem with solar towers is that each installation requires large areas for the thousands of mirrors required for focusing the sunlight onto the top of the tower. This limits the suitability of solar towers to desert areas, or places where there are wide open spaces.

An unforeseen problem when Ivahpah was built, is that birds are killed when their feathers catch fire as they fly through the concentrated sun rays.

A second type of concentrating solar power plant uses solar troughs. See Figure 9.

Curved mirrors focus sunlight onto a tube containing a fluid that runs the length of the mirrors, of which there are many rows.

The fluid is used in a heat exchanger, such as a boiler containing water, to produce steam, that drives a turbine generator to produce electricity.

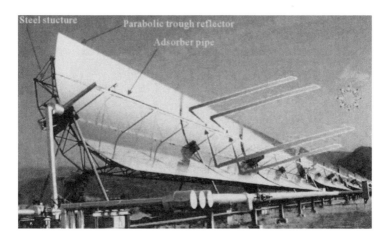

Figure 9. CSP Solar Trough Photo from NREL

In an effort to increase the amount of electricity generated by tower and trough type CSP plants, some have included salt heat-sinks in which to store heat so that electricity can be generated for a few hours after the sun sets.

Another type of CSP solar power plant uses a parabolic receiver to focus sunlight onto a receiver that uses a gas to drive a sterling engine to generate electricity, Figure 10.

These units are located in groups, similar to a wind farm, but require far less area.

Utility scale PV solar installations have also been built.

PV solar installations use PV panels arrayed across large land areas, with or without the ability to track the sun. Agua Caliente is one such PV power plant.

Figure 10. Photo CSP Parabolic Unit

It's also possible to build CSP and PV solar installations to operate in conjunction with natural gas power plants.

An example is Florida Power and Light's Martin power plant where 51 linear miles of solar troughs were installed to augment the capacity of an existing natural gas power plant.

Concentrating Solar Energy

All solar installations can essentially only generate electricity when the sun shines, so, like wind, they are unreliable.

Clouds will interrupt the generation of electricity, either intermittently, or completely, as on stormy days.

Electricity from all types of CSP installations is also extremely expensive, at approximately five times the cost of generating electricity with natural gas combine cycle power plants.[17]

Chapter 8

PV Rooftop Solar Energy

The combined size of PV rooftop solar installations at the end of 2014 was estimated to be 6,200 MW by the Solar Energy Industries Association, but no one really knows.

Even so, the main impetus behind installing PV rooftop solar systems has been government mandates and subsidies.

For the United States, there are very few areas where PV rooftop solar is economically viable. For nearly all areas in the United States subsidies are required to entice home owners and businesses to install PV rooftop solar.

PV rooftop solar panels can produce 0.75 kWh of electricity per square yard of panel per day.[18]

A two-story, 3,000-square-foot home will have a total roof top area of approximately 1,500 square feet. But, since only half can face the sun, the available area is 750 sq. feet.

With residential electricity costing 12 cents per kWh in Phoenix Arizona, this installation can save $6.88 every day the sun shines. If the sun shone 365 days every year in every city in America, there might be some small justification for investing in PV rooftop solar.

However, the sun does not shine every day and this is one reason why the economics are bad.

In Phoenix, Arizona, where the sun shines 211 days each year, an installation on a two-story, 3,000-sq. ft. home would reduce the homeowners electric utility bill by $1,583.

PV Rooftop Solar Energy

It's true, there would be partly sunny days that might improve the picture, but offsetting that possible benefit is that few homes can have their roofs aimed directly at the sun, and this would reduce the efficiency of the rooftop PV panels. (Equipping rooftop panels to follow the sun during the day and as the sun travels north and south during the seasons would substantially increase costs.)

Even with panels now costing half of what they did a few years ago, installing them makes no economic sense.

The economics don't favor PV rooftop solar.

Many Internet sites have information on PV rooftop solar installations.

Some offer to provide a free estimate of the savings people can achieve with a PV rooftop solar system, and some have an online calculator allowing the viewer to immediately determine possible savings.

Just be careful, the calculator can distort the payback.

In one ploy, the cost of the PV rooftop solar system is added to the value of the home.

But a PV rooftop solar system is probably similar to a swimming pool. The homeowner rarely recovers the cost of the swimming pool when selling the home.

These calculators frequently don't include the cost of installation, which can be substantial and reduce the calculated return on investment.

In addition, the calculator deducts the federal 30% tax credit from the investment, so the payback is determined using the net cost after the tax credit.

While the homeowner benefits from the tax credit, other homeowners are subsidizing the system's cost with tax payer money.

If PV rooftop systems are so good, why do sellers of these systems have to distort their true value?

The answer is simple.

In most locations, PV rooftop solar systems aren't a good investment.

They require a subsidy to make a bad investment appear good.

Online calculators also use a variable to reflect the location of the installation. The variable is likely to be the insolation value for that location.

Insolation values are determined for locations around the world, and show how much sunlight falls on the Earth at each location. It's frequently expressed as kWh/square meter/day.

These values vary widely. For example:

- Central Australia = 5.89 kWh/m2/day
- Helsinki, Finland = 2.41 kWh/m2/day

And in the United States, see Figure 10:

- Phoenix, Arizona = 5.38 kWh/m2/day
- Minneapolis, Minnesota = 3.68 kWh/m2/day

PV Rooftop Solar Energy

One calculator estimated in July 2015, that a PV rooftop solar installation in Albany NY, using 1,500 kWh of electricity monthly, would have a payback of 12 years, without tax credits, assuming the rooftop system faces south.

If the installation doesn't face due south, the amount of electricity generated by the installation is reduced, and the return on investment worsened.

An installation in Albany that faces east or west, could require an additional 2 years to recover the investment.

From an investment perspective, a 12-year payback is abysmal, and made worse if the installation doesn't face due south.

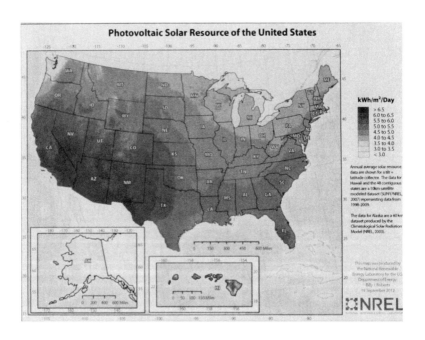

Figure 11. Map of U.S. Insolation Values

PV Rooftop Solar Energy

Payback periods at other locations vary depending on insolation levels, and could be much longer

Here is a sampling of payback periods, without tax credits, as determined by one of the Internet calculators in July 2015:

- Atlanta, GA, 15 years
- Chicago, IL, 25 years
- Lincoln, NE, 17 years
- Pittsburgh, PA, 21 years
- Spokane, WA, 22 years
- Tampa, FL, 11 years
- Tucson, AZ, 10 years

Appendix 1 provides payback periods from the same calculator, without using tax credits, for typical locations in every state.

It's obvious there are few places in the United States where paybacks without subsidies are reasonable, say less than 5 years.

Note the payback for an installation in Tucson, Arizona, one of the more favorable locations in the United States, is 10 years.

In addition, PV rooftop solar panels are only expected to last 20 years, so in many locations the panels would be scrapped before they had paid for themselves.

Because of the variation in insolation values and the direction PV rooftop systems face, anyone enticed into buying a PV rooftop solar system should insist on a warranty, where, if the savings over a year aren't achieved, the company making the installation would rebate a proportionate portion of the installation cost.

For example, if the saving were 10% less than guaranteed, there would be a rebate equal to 10% of the installed cost.

This only makes good business sense.

The cost of PV rooftop solar will probably come down over the next decade. But even if the panels cost half of what they do today, PV rooftop solar would still be a bad investment throughout nearly all the United States.

While PV rooftop solar is a bad investment for most homeowners, some PV solar installation companies are leasing PV rooftop installations to homeowners, which turns a bad investment into a potentially good proposition for the homeowner.

Leasing the PV rooftop system to homeowners, where homeowners pay less for the monthly lease than they would previously have paid the utility, makes the lease attractive to homeowners.

The PV leasing company then uses the federal and state subsidies, and net metering payments from the utility, along with the usual depreciation charges, to generate a profit.

Net metering usually results in the utility not being compensated for the installation and maintenance of transmission and distribution lines, which isn't fair to the utility and its share holders, many of which are pension funds.[19]

The leasing company is happy because it makes a profit, homeowners are happy because thy cut their cost for electricity, but taxpayers should be unhappy because it's their tax money being used to pay for the subsidies.

Utilities are unhappy because they get stuck with paying for the maintenance of transmission and distribution lines, and because renewables, wind and solar, are threatening the grid.

Chapter 9 explains why utilities are threatened, and why everyone should be concerned by threats to the survival of utility companies.

PV rooftop solar contracts from third parties may contain escalation provisions for amounts paid by homeowners to third party leasing companies that assume electricity rates will increase over the life of the contract. Homeowners should be careful that these are tied to actual rate increases.

Homeowners should also be careful about other contractual provisions in leases, such as liability, and should be certain their homeowner insurance rates won't increase.

A few states prevent homeowners from entering into leasing agreements with companies that install PV rooftop solar systems.

Groups promoting solar within those states claim it's unfair to deny homeowners the right to enter into such agreements.

PV rooftop systems are expensive, and require an upfront investment that can range from $20,000 to $30,000, or more, depending on the size of the home and available rooftop area.

Adding batteries, that cost around $7,000 each, to allow going off grid or allow for demand response when electricity demand is at its peak, can add substantially to the required investment.

PV Rooftop Solar Energy

As shown above, installing rooftop systems in most parts of the United States is uneconomic without subsides.

The crux of the matter is that leasing companies turn a bad economic investment into an attractive proposition for homeowners.

While the current number of PV rooftop systems is small, they have little effect on the grid, but widespread usage of PV rooftop systems would put the nationwide grid at risk.

The grid is essential for the country to function.

It's not only homeowners who use the grid, it's industry, commercial centers and, often overlooked, people who live in cities.

For example, how can electricity generated by wind in Montana get to the people who live in cities without the grid?

Going off grid is not a viable option for most industries, or for nearly all commercial centers. And it's rarely an option for homeowners.

Articles at Energy Matters, by Roger Andrews, makes it clear that the U.S. grid is absolutely essential.[20]

Mr. Andrews' analyses demonstrate that PV rooftop solar will not allow, with any reasonable assumptions regarding cost, for homeowners to go off grid, except in the Southern states.

If renewables must be dispatched first, even when they cost more than electricity generated from conventional sources, and PV rooftop installations are allowed to proliferate with utilities not being sufficiently compensated for maintaining transmission and distribution lines, utilities will go bankrupt.

In the Northwest, the Bonneville Power Administration must dispatch expensive wind generated electricity and flush water, through or around its dams, bypassing hydro generators that could generate electricity far more cheaply, in order to comply with rulings requiring the dispatch of wind energy first.[21]

Dispatching renewables first is the rule in Germany, and Germany is the poster child of how utilities are being threatened by renewables.

The largest German utility, E.ON, has decided to divest all its fossil fuel generating facilities and focus entirely on renewables.

The second largest utility, RWE, has also undertaken to dispose of its fossil fuel assets.

Since it's unlikely that anyone would buy a business losing money, these actions will likely result in the nationalization of all the fossil fuel assets and transmission lines so as to keep the lights on in Germany.

Germany has undertaken its energiewende program so as to cut CO_2 emissions.

At the end of 2014, Germany had cut its CO_2 emissions by only 27% from 1990 levels, with an objective of cutting them 80%, and yet, Germany has already run into serious problems.

The examples of high prices for electricity and potential reliability issues in Germany should be cautionary tales for the United States.

PV Rooftop Solar Energy

Unreliable solar energy drives up the cost of electricity, and deprives Americans and American industry access to the least costly, most reliable electricity, which is generated using fossil fuels.

Using subsidies to promote the widespread use of PV rooftop solar threatens the survivability of both the grid and the electric utility industry.

Chapter 9

The Utility Death Spiral

The past three chapters on wind and solar serve as an introduction to the current chapter which explains why renewables create the utility death spiral.

The term that's currently de rigueur, is "over-generation."

It's supposed to indicate that power generation, using fossil fuels, is being done in excess.

What over-generation really means is the "over-generation" of expensive and unreliable electricity by wind and solar.

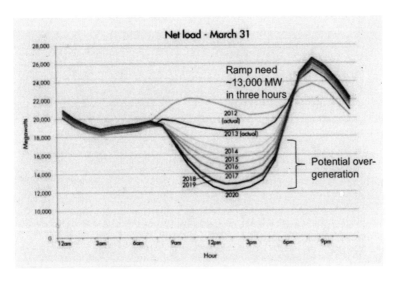

Diagram 1, CAISO Duck, Net Load

The Utility Death Spiral

The Duck curve was created by the California Independent System Operator (CAISO) because California is confronted with too much electricity being generated by wind and solar, resulting in traditional power generation being underutilized. The curve is for a single day in March, but curves for the rest of the year are similar.

The year 2020, in Diagrams 1 & 2 represents 33% penetration by renewables.

The very top curve depicts when virtually all of the electricity was generated by conventional sources in 2012. Conventional sources being primarily fossil fuels. In succeeding years, 2013 through 2020, the belly of the Duck depicts the increasing effect of electricity generated by renewables, primarily wind and solar.

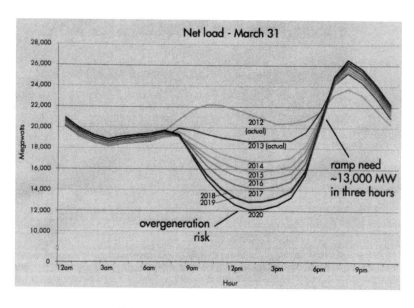

Diagram 2, Over Generation

The Utility Death Spiral

The heavy dark lines in Diagram 1 & 2 depict the electricity supplied by conventional sources in 2020.

The difference between the topmost curve and the bottom curve represents the supposed over-generation from traditional sources, but it more accurately can be defined as the over-generation by expensive and unreliable renewables.

The diagrams reflect net load which is calculated by taking the forecasted load and subtracting the forecasted electricity production from variable generation resources, wind and solar.

The belly of the Duck becomes extended in the years beyond 2020, and absolutely bloated by 2030, when California expects renewables to achieve 50% penetration.

Until early morning, in Diagram 2, a certain amount of electricity is supplied to the system from conventional fossil fuels. But during the day, only about half of the available capacity of conventional sources is used in 2020, with perhaps only one-third, or less, of the available fossil fuel capacity being used in 2030, or later.

And then there is the sudden shock in the evening when the system must suddenly ramp up 13,000 MW within three hours to its highest capacity utilization, at around 25,000 MW, placing a terrible strain on the system.

The stresses incurred are actually worse, because there is additional ramping up and down of traditional power plants due to the sudden loss of electricity when the wind stops blowing or the sun stops shining at other times during the day.

Sudden ramping up and down causes severe thermal and other stresses on the system, including equipment damage due to differences in thermal expansion between the materials used in boilers, gas turbines and other equipment.

Germany is already rapidly approaching the situation in which California will find itself in 2020. In Germany, the penetration in 2014 of renewables, i.e., wind, solar, geothermal and hydro, was 21%, and increasing rapidly.[22]

In Germany the utilities are trying to dispose of all their traditional methods of power generation, because they can't survive with the bloated belly of the Duck.

German citizens are already paying four to five times as much for electricity as do Americans.

Imagine the bloated belly of the Duck extending far lower than shown in Diagram 2, perhaps to zero.

The Duck makes abundantly clear the impossible plight of utilities that supply electricity from conventional sources, while also showing why California can't survive without electricity from fossil fuels.

If the belly of the Duck is distended far lower than shown in the diagrams, the utilities in California won't be able to survive when they can only generate 1/3, or less, of the electricity for which they have built capacity.

If utilities have invested in and built 50,000 MW of conventional fossil fuel capacity, but are only allowed to generate enough electricity from conventional sources to cover 1/2 their investment, one of two things must happen.

The Utility Death Spiral

Either:

1. The utilities must charge much higher rates than their current rates.

2. The government must take over, i.e., expropriate, the utilities and use tax payer money to subsidize the generation of electricity from fossil fuels, using, what will then be power plants and grid owned by the government.

The eventual takeover of the utility industry by the government is the end result of the utility death spiral.

Perhaps an analogy would make this clear.

Suppose you had built a restaurant to serve 200 people, but later the government decrees that half the restaurant must be used by people who bring their own food.

You needed to serve meals to 200 paying customers in order to cover your investment and make a small profit, but with half your investment being used by people who don't pay for food and services you only receive half the income needed to cover your investment.

In this scenario, you go bankrupt.

This is the situation faced by utilities when renewables, especially PV rooftop solar, displaces half or more of the electricity the utilities can produce from their investment in generating capacity … generating capacity that must be maintained to meet peak demand.

The Utility Death Spiral

And utilities can't remove any of their investment because they must be able to meet peak demand when the sun stops shining and the wind stops blowing.

With the unrestricted growth of renewables the end result will be that people and industry will be paying:

1. For expensive and unreliable electricity generated by wind and solar.

2. While also paying much higher rates, either directly to the utilities to cover their underutilized investments, or through higher taxes for electricity generated by government owned utilities using traditional fossil fuels to meet peak demand.

The California government hopes that storage can minimize the Duck's negative effects. But, except for pumped storage, which can't be built because it requires dams that extreme environmentalists oppose, the prospects for sufficient storage to offset the negative aspects of the duck are not promising. Especially if we consider the entire country.

Even if storage is developed, it is more expensive than building natural gas power plants, so the consumer will still be paying much more for a less reliable system.

The Duck tells all, and his message isn't good for America if the current infatuation with renewables continues.

Over-generation is caused by inefficient, expensive and unreliable renewables.

The Utility Death Spiral

This isn't the AFLAC Duck paying money to people who have been injured.

This is the opposite, where the California Duck causes people to pay a great deal more for their electricity, and for a system that is less reliable.

While the Duck curve displays the over generation caused by renewables, the effect of renewables can better be seen by examining a load curve. Load curves actually vary by time of year and location, however the generalized load curve depicted in Diagram 3 shows the theoretical impact renewables will have on utility revenues ... and the vital need for storage.

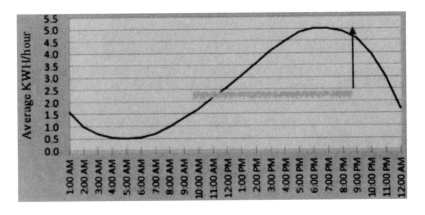

Diagram 3. Load Curve

The curve depicts the amount of electricity used hourly during the day. It can be referred to as a demand or load curve.

The Utility Death Spiral

During the night and after sunset, the load is mostly supplied by base load, dispatchable power generation sources, primarily from coal-fired or natural gas combined cycle power plants, though it could also be from hydroelectric or geothermal sources.

At some point during the morning, wind and solar generated electricity replace the electricity generated by base load power plants.

Conceptually, wind and solar displace electricity generated by fossil fuels during the day, so the area above the horizontal line is no longer served by electricity generated from fossil fuels. It's served by wind and solar.

When the sun sets in the evening, wind and solar, but especially solar, stop generating electricity, and demand must be met by base load power plants, primarily by power plants using fossil fuels.

The vertical line depicts the sudden ramping up of fossil fuel generation assets as the sun sets and wind and solar no longer provide large amounts of electricity.

Theoretically, storage of electricity would extend the supply of electricity from renewable sources for a period of time, and have the effect of minimizing, or theoretically, eliminating, the sudden ramp up depicted by the vertical line. While storage may solve the need to suddenly ramp up fossil fuel power plants, it also further reduces the electricity sold by the utility from fossil fuel power plants.

Except for pumped storage, batteries provide the only realistic technology for large amounts of storage today. Storage by hydrogen, methane, compressed air (CAES) or thermal techniques are, at best, questionable ... and probably unlikely in sufficiently large quantities.

Germany's energiewende policy provides a real world example of what happens when renewables displace electricity that has traditionally been supplied by fossil fuel power plants.

In 2014, on average, less than 20% of Germany's load was supplied by wind and solar.

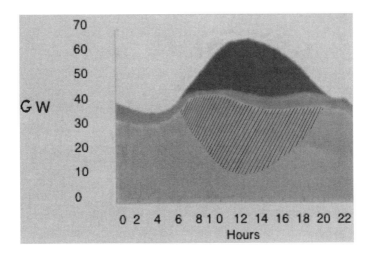

Diagram 4. German Load Curve from Citi[23]

The top of the curve shows the load during a summer day, peaking at around 65 GW at midday. The load supplied by solar, is the area in grey at the middle part of the diagram, while the load supplied by wind is the dark grey area at the top.

The Utility Death Spiral

The area in light grey is the load supplied primarily by fossil fuels, with some nuclear included. Nuclear is being phased out.

The hatched area is a rough depiction of the encroachment of renewables on fossil fuel generation if the percentage of renewables approaches 80%, as forecast by Germany's energiewende policy. The hatched area represents additional revenues lost to utilities from their fossil fuel generation assets.

The more renewables, the less revenue for utilities from their fossil fuel assets.

Renewables are displacing fossil fuel generated electricity produced by utilities, which is why E.ON and RWE, two of Germany's largest utilities, are tying to divest themselves of all their fossil fuel power generation assets.

While it's impossible to predict the future, there appear to be only two possible outcomes in Germany. Either:

1. Electricity rates being charged to consumers, who already pay 4 to 5 times as much as do Americans, are increased substantially above today's rates to provide a capacity fee for utilities to maintain their fossil fuel assets, or

2. The government nationalizes the electric utility industry and the grid, so that tax payer money is used to operate the uneconomic utility industry.

In the United States, California is leading the charge in mandating the use of large quantities of renewables. Elsewhere, 31 states have renewable portfolio standards (RPS) mandating that as much as 25% of electricity be from renewables.

The Utility Death Spiral

In the United States, most states, including California, do not consider hydro to be renewable.

Perhaps it is time to consider why states are mandating the use of renewables?

It can't be to lower costs to consumers, because wind and solar are both more expensive than generating electricity using natural gas or coal.

The Energy Information Administration (EIA) predicts that in 2020 the cost of electricity from most renewable sources will still be higher than for natural gas[24]:

- On-shore wind, 7.3 cents per kWh
- Off-shore wind, 19.7 cents per kWh
- PV solar, 12.5 cents per kWh
- Thermal solar, 24 cents per kWh

The EIA estimates that electricity produced by natural gas will be 7.5 cents per kWh in 2020. (This estimate of natural gas generation cost is probably too high given the surplus of natural gas that will keep prices low.)

The current cost of wind and all forms of solar is significantly higher than the costs predicted for 2020.

In other words, renewables are, and will remain, more costly than electricity generated by fossil fuels, even without adding the cost of backup generation for when the wind stops blowing and the sun stops shining, or the cost of transmitting wind generated electricity long distances.

The Utility Death Spiral

The reason renewables are being forced onto the system is because they don't emit CO2.

Elon Musk, in his press conference and video introducing Tesla's Powerwall battery, said batteries and solar would prevent catastrophic climate change by reducing CO2 emissions.[25]

In other words, the utility industry is being threatened, and costs to consumers are being dramatically increased because of the government's efforts to cut CO2 emissions.

The Duck curve, together with Germany's real world experience, prove that renewables are not beneficial. Renewables are expensive and unreliable, and have the potential to destroy the utility industry, causing consumers and industry to pay much more for electricity, which deprives the economy of the benefits of additional consumer spending and investment by industry.

Worst of all, renewables could result in the government taking over the utility industry, with government bureaucrats running the industry.

Blindly replacing fossil fuels with renewables, without evaluating their relative costs or impact on the economy, is not in our best interests.

Environmental organizations have vigorously opposed nuclear power, so it is not addressed here. Information in the appendix explains why nuclear power may be declining in the United States as the result of the misguided attacks against it.

While renewables limit and restrict the use of energy, because of their high cost, unreliability and potential need for a government takeover of the industry, fossil fuels provide the means for mitigating any of the possible effects of climate change.

Chapter 10

Inadequacy of Biofuels

Many attempts have been made to develop biofuels to replace liquid transportation fossil fuels, primarily gasoline, diesel and jet fuel.

While some of these efforts have worked in the laboratory or in trials, most, if not all, have yet to advance beyond the pilot plant stage.

Because of the diverse attempts to develop biofuels, only three companies will be examined in this chapter. Each company, however, uses a different feedstock for producing their biofuels.

So as not to detract in any way from their efforts or inhibit their potential success, the companies will not be named. Actual data will be taken from their websites and it will be used to examine whether the efforts to cut CO_2 emissions are realistic.

While processes may work, the resulting product may be too expensive or there may be inadequate resources to support the output required to replace the fossil fuel in question.

Company X uses wood as its feedstock.
Company Y uses garbage as its feedstock.
Company Z uses algae as its feedstock.
Companies X and Y are attempting to produce a biofuel that can replace existing jet fuel.

Jet fuel made from oil emits CO_2, and, according to the UNFCCC, IPCC and EPA, CO_2 emissions cause climate change.

Inadequacy of Biofuels

Aviation CO2 emissions are only 2% of total worldwide CO2 emissions, while China currently accounts for approximately 30% of worldwide CO2 emissions.

China's emissions are forecast to increase approximately 170% by 2030 as the result of the recent Obama - China agreement, as shown in Figure 6. This increase will far outweigh any possible reduction in CO2 emissions from forcing the aviation industry to switch to biofuels.

If the UNFCCC and EPA enact regulations prohibiting the use of traditional jet fuel, it would cause serious problems for the airline industry which expects the number of airline passengers to more than double by 2035.

The FAA said it will award $7.7 million in contracts to eight companies to help develop biofuels from sources such as alcohols, sugars, biomass, and organic materials known as pyrolysis oils.

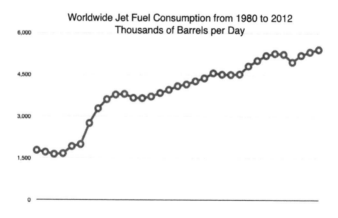

Figure 12. Jet Fuel Daily Consumption, bbls/day, from EIA[26]

Consumption in 2012 was 5,418,000 barrels per day.

Inadequacy of Biofuels

The Federal Aviation Administration (FAA) and its German counterpart, the BMVBS, signed an agreement in 2014 to promote, develop and use jet biofuels.

Based on daily consumption, as shown in Figure 12, nearly 2 billion barrels of jet fuel were used in 2012.

Nearly 116 billion gallons of jet biofuel will be needed annually by 2030, based on a doubling of miles flown and assuming the efficiency of jet engines is improved by 30%.

This is the amount of jet biofuel that must be produced if traditional jet fuel is outlawed.

Will it be possible to produce this much jet biofuel economically, from available feedstocks?

Company X

Company X's website states it can produce 16 million gallons of jet biofuel from 175,000 tons of woody pulp annually. The Clemson Extension provides information on tons per acre for various ages and heights of pine trees.[27] For example, one acre of 70-ft-tall, 30-year-old loblolly pine trees can produce 140 tons of pulp.

As a result, 9.1 million acres of 30 year old pine trees are required each year as feedstock for Company X's process to produce 116 billion gallons of jet fuel, the amount required by the world in 2030.[28] This is larger than the area of New Hampshire.

In other words, an area greater than thirty times the size of New Hampshire, or 262 million acres, is required by Company X to grow enough trees to meet the annual requirement for jet biofuel where trees are harvested annually.

Inadequacy of Biofuels

There are currently 190 million acres of forests in the United States, so this would have to be a worldwide effort.

The effort to plant 9.1 million acres each year over the next 14 years, between now and 2030, would have to begin immediately, and harvesting of existing trees would have to be done for 16 years before the new crop of trees would be available.

While all this sounds possible, it defies logic.

<u>Company Y</u>

Company Y's website says it can produce 10 million gallons of jet biofuel annually with a 200,000 ton supply of municipal solid waste (MSW) at its new pilot plant. To supply 116 billion gallons of jet biofuel, the company will require 2.3 billion tons of MSW with many additional plants.

The average American produces approximately 2.3 pounds of residential MSW daily, which, for 350 million Americans, amounts to 143 million tons annually.

In other words, the company will need 16 times as much residential MSW as is generated annually in the United States.

It should be noted that Company Y's website provides alternative data. This data indicates a smaller amount of biofuel would be produced from 1.3 billion tons of MSW generated worldwide. This data indicates an output of 32 gallons per ton of garbage, while the pilot plant indicates an output of 50 gallons per ton of garbage.

Company Y also states that output will include both diesel and jet biofuel.

Inadequacy of Biofuels

Regardless of the actual output per ton, it's obvious that there is insufficient feedstock, i.e., MSW, to provide all the required jet biofuel if traditional jet fuel is outlawed.

<u>Company Z</u>

Company Z has an algae farm in a western state where it claims to be developing oil from algae. Facts from published information are limited, though a recent study made headlines.

Here is what Company Z's website says:

- The process can produce 1,000,000 gallons of a drop-in oil substitute per year, on 300 acres
- The product will emit 60 to 70% less CO_2 than petroleum fuels
- Large quantities of non-potable saltwater are used in the process

There is no mention of how much it costs to produce a barrel of oil.

It also appears as though large amounts of sunlight is needed, and that the plant be located on a gradual incline so that gravity can be used for the required flow of water.

These requirements limit where the process can be used.

The website also says a 5,000-barrel-per-day plant will be built and the results known from that site in 2018.

Here's what we can extrapolate from what we know, excluding any estimate of whether the oil, without subsidies, will be competitive.

Inadequacy of Biofuels

- The United States, according to the EIA, used 3.26 billion barrels of oil in 2014 for gasoline

- It would require about 6,418 square miles on which to produce 326,000,000 barrels of oil per year, or 10% of the oil used each year in the United States for gasoline

- For comparison, Connecticut has an area of 5,544 sq. miles

While the state of Connecticut can fit nicely into Arizona, New Mexico, Nevada or Texas, many times over, much of these areas may not meet the requirements for water and gentle topography required for algae farms.

Nevada, for example, has large unused land areas owned by the Interior Department, but doesn't have water.

Areas of Texas have brackish water available, but the land is owned by people, and they may not want to sell, or the land may be expensive and increase the cost of the oil produced from algae. For example, the Permian Basin, an area rich in oil deposits, covers much of Texas.

Producing all of the gasoline used by the United States would require an area larger than Georgia.

On the surface, based on this limited information, it would appear that Company Y may be able to produce some oil.

Whether it can ever replace large quantities of competitively priced gasoline remains to be seen.

The issue with algae isn't whether it will work, because it can.

Inadequacy of Biofuels

The issue is whether algae derived fuels can be produced in sufficient quantity to replace large quantities of fossil fuels, and whether it can be produced at a price that's competitive with gasoline or diesel fuel.

Currently, the answer for algae is no.

This brief overview of three biofuel companies can only provide a glimpse into the practicality of replacing fossil fuels with biofuels.

In addition, the cost of the 450,000 gallons of jet biofuels produced for the U.S. Navy for use in some exercises was six times the cost of traditional jet fuel.[29]

Only three feedstocks, trees, garbage and algae, were discussed in this chapter as potential feedstocks for producing biofuels, but, other than food crops, there are not many additional feedstocks capable of producing biofuels having the required energy content.

Using a food crop has been called a crime against humanity, and should be avoided for that reason.

There may be other biofuel ideas in the laboratory, but this review of three important alternative feedstocks indicates it will be very difficult for biofuels to be able to supply the required amounts of gasoline, diesel fuel or jet fuel.

The possibility of producing biofuels economically and in required quantities seems remote … if not absurd.

Chapter 11

Role of Taxpayer Funded Subsidies

Subsidies have played a huge role in the development of wind and solar installations.

Perhaps the biggest untruth with respect to energy is that fossil fuels receive huge subsidies, while clean energy, such as wind and solar, receive few subsidies.

Even if true, which it isn't, fossil fuel subsidies, to the extent they exist, are for increasing the availability of energy, which benefits Americans, while clean energy subsidies harm Americans with higher taxes and higher costs for electricity and gasoline.

While this infers there are good and bad subsidies, there probably are few, if any, good subsidies since they distort free markets.

One of the components of this Big Untruth is that the United Sates spends $10 to $500 billion annually to defend oil interests overseas. Or as the web site *Oil Change International* says, "Political destabilization and lives lost due to military force in the [Mideast] region — casualties of the insatiable U.S. thirst for oil."

This is obviously a political statement rather than an objective one, so let's ignore it for the time being, though it does lend emotional support, not credence, to the Big Untruth.

Besides, we are producing more of the oil we use, and in combination with Canada, can probably produce all the oil we need.

But what are fossil fuel subsidies?

Role of Taxpayer Funded Subsidies

The IEA says, "[Our] latest estimates indicate that fossil-fuel consumption subsidies worldwide amounted to $409 billion in 2010," Figure13.

Fossil fuel consumption subsidies are discounts given by oil producing countries, in the form of discounted gasoline prices, to their citizens.

These are not subsidies helping oil companies.

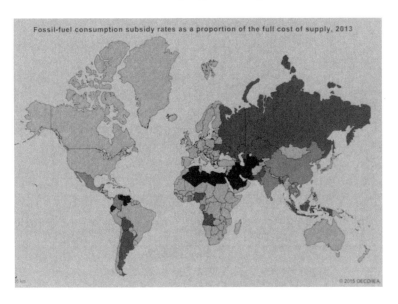

Figure 13. IEA Subsidy Chart: Black, Subsidy over 50%, Dark Grey, 20% to 50%

This is an important distinction and fact. The preponderance of subsidies, as reported by the IEA, are for people in oil producing countries. They are consumption subsidies to help lower the cost of living for the citizens of oil producing countries such as Venezuela and Saudi Arabia.

They are not subsidies to support the oil industry.

Unfortunately, the media includes these consumption subsidies when they accuse the fossil fuel industry of being subsidized.

So, what about subsidies for fossil fuel companies? Or for renewables?

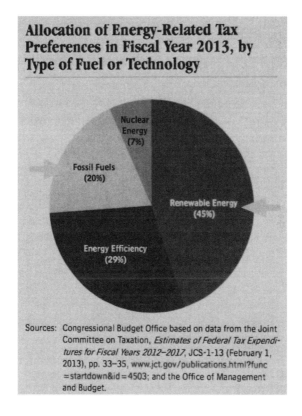

Figure 14. Energy Tax Preferences

89

Role of Taxpayer Funded Subsidies

A report providing data on subsidies was produced by the Congressional Budget Office on March 13, 2013.

Testimony given to the House of Representatives included Figure 14 that provides a clearer picture of where tax preferences, i.e., subsidies, are applied.

Fossil fuels received 20%, while renewable energy received 45% of the subsidies, despite renewables, other than hydro, providing less than 6% of the energy consumed.[30]

Table 3		
US Primary Energy Consumption by Source		
	Quadrillion BTU	Percent
Petroleum	34.8	35%
Natural gas	27.6	28%
Coal	18	18%
Nuclear	8.3	8%
Biomass	4.8	4%
Hydro	2.5	2.5%
Wind	1.7	1.7%
Solar	0.4	<1%
Geothermal	0.2	<1%
Total	98.3	
EIA, Monthly Energy Review, Table 1.3		
Note: Does not include PV Rooftop Solar		

Role of Taxpayer Funded Subsidies

This isn't the entire picture however, because it doesn't include loans and grants, such as the one given to Solyndra. The report says, "[Direct support and loan guarantees] amounted to $3.4 billion in both 2012 and 2013. About half of that support is directed toward energy efficiency and renewable energy in 2013."

The report also provides annual data on fossil fuel and renewable energy subsidies from 1977 to 2013, as shown in Figure 15.

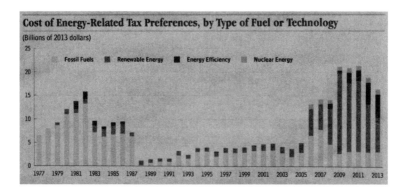

Figure 15. Energy Tax Preferences 1972 - 2013
Light grey: Fossil fuel. Dark grey: Renewables. Black: Energy efficiency.

From 1977 until 1987, fossil fuels received large subsidies in response to the Arab oil embargo. They represented efforts by the United States to achieve energy independence. It was during this period that there were long lines at gasoline stations. President Carter supported a $5 billion effort to mine shale rock in Colorado, where oil could be extracted from the rock by crushing and cooking it.

After these initial years, and until 2006, there were far fewer subsidies for either fossil fuels or renewables.

Beginning in 2006, subsidies increased substantially, and from 2008 to the present, renewables have received far larger subsidies than fossil fuels.

This puts to rest the idea that wind and solar should get subsidies, because they already do, and the subsidies renewables get are very large ... much, much larger than are received by fossil fuels.

Some extremists seek to include other items as subsidies, or they want the accounting practices changed to favor renewables.

Generally accepted accounting principles (GAAP) are in place to provide accurate and consistent information to investors and others, and to prevent the unscrupulous from distorting the facts.

A good example is the depletion allowance. The depletion allowance provides funds for exploration and development of new resources to replace those being removed by drilling or mining.

This is analogous to depreciation that provides funds to replace equipment that wears out.

Table 4 attempts to place some of these arguments in perspective, but the testimony by the Congressional Budget Office provides a more complete analysis.

Role of Taxpayer Funded Subsidies

Table 4		
Tax Preference	**Fossil Fuel Company**	**Wind Energy Company**
Depreciation of equipment, buildings and structures	Yes	Yes
Depletion	Yes & No[a]	No
Production tax credit	No	Yes
Expensing drilling costs	Yes	No
Foreign Tax credit	Yes	No[b]
a. Not available to large, integrated companies. b. Only if domestic company operated internationally.		

Referring to Table 3 and Figure 15, based on the amount of energy produced per dollar of subsidy, renewables are currently receiving subsidies that are over 40 times larger than subsidies given to fossil fuels.[31]

Without subsidies, it's doubtful that wind and solar would be utilized except in remote areas where the grid is not available, or on islands where oil, diesel fuel or natural gas must be imported.

Electricity from wind and solar are more expensive than electricity generated by natural gas or coal.

Role of Taxpayer Funded Subsidies

Wind and solar are being promoted and given large taxpayer funded subsidies, way out of proportion to the amount of energy they produce, because they cut CO2 emissions, even though wind and solar are uneconomic.

Part Three

Reality

Reality bites... and doesn't let go.

~Author Unknown

Chapter 12

Carbon Capture & Sequestration

Carbon capture and sequestration (CCS), in conjunction with so-called clean coal technology for generating electricity, has been viewed as the enabling technology for dramatically cutting CO_2 emissions.

CCS assumes that all, or nearly all, CO_2 emissions from power plants can be captured and then injected underground, i.e., sequestered, to prevent the CO_2 from reaching the atmosphere.

This is a very appealing idea because it means that existing coal-fired and natural gas combined cycle (NGCC) power plants can remain in operation. It also would mean that new fossil fuel power plants could be built where the CO_2 is captured and sequestered underground.

There are two fundamental problems with CCS that make the concept unworkable.

First, there is the problem of capturing CO_2.

Second, there is the problem of sequestering CO_2 underground, with absolute certainty that it will never escape to the atmosphere, at least for thousands of years.

The initial hurdle is to develop methods for capturing CO_2 from existing power plants, and several methods have been proposed and tried.

FutureGen, a pilot project located in Illinois, begun under president George W. Bush in 2003, was supposed to prove that capturing CO2 could work, but the project has recently been cancelled.

Even so, there is enough information available to believe that carbon capture might be possible.

The main problem with carbon capture from existing power plants is that the parasitic losses require the power plant to be derated 30 to 40%. In other words, a 300 MW power plant would have to use 30 to 40% of the electricity it generates to capture and compress the CO2 it produced, which would result in the plant only being capable of producing 200 MW of electricity for the grid.

Total installed U.S. generating capacity using fossil fuels amounted to 780,000 MW in 2012. If all these were equipped to capture CO2, an additional 260,000 MW of capacity would have to be built to replace the capacity lost due to reconfiguring the plants to capture CO2.

At 1,000 MW per power plant, 260 new power plants would have to be built at a cost of approximately $350 billion, not counting the billions required to modify the existing power plants to capture CO2.

If all 260 plants were built as NGCC plants, they would have to be around 40% larger and more expensive than those being built today to accommodate carbon capture.

Another type of power plant has also been proposed for capturing CO2, known as an integrated gasification combined cycle (IGCC) power plant.

Carbon Capture & Sequestration

IGCC power plants essentially cook the coal to create a syngas, composed mostly of CO_2, hydrogen and CO. The CO_2 is then extracted from the syngas and the remaining gasses, mostly hydrogen, are burned in a gas turbine to generate electricity.

Three IGCC plants have been built, each at a cost of nearly $6,000 / KW, which is about what a new nuclear power plant costs.

However, most have been built without the ability to capture CO_2. For comparison, a new NGCC power plant without the ability to capture CO_2 costs approximately $1,100 / KW.

The three IGCC plants built in the United States are: The first by Tampa Electric, a second by Duke at Edwardsport, Indiana, and the third, which is under construction in Kemper County, Mississippi, by The Southern Company.

The Kemper plant is now projected to cost $6.2 billion, and a partner of the Southern Company, South Mississippi Electric Power Association, has backed out of the project because of its high cost.

Costs were to have been capped at $2.9 billion, but with a cost of $6.2 billion, and still possibly rising, rate payers in Mississippi may see a very large increase in what they pay for electricity.

IGCC power plants have been an economic failure.

Once the CO_2 has been captured, it must be compressed to around 2,000 psi and transported to where it can be injected into a geologic formation underground.

The Pacific Northwest National Laboratory has estimated it would require building around 20,000 miles of new pipelines to transport the CO_2.[32]

Carbon Capture & Sequestration

Collectively, CO2 pipelines would be over thirteen times longer than the Trans Alaska Pipeline System (TAPS) which stretches 800 miles from Prudhoe Bay to Valdez.

Some people point to the fact that small quantities of CO2 have been used for enhanced oil recovery (EOR), and that a few pipelines have ben built for this purpose. But the quantities involved have been tiny compared to what would be required if sequestration was to be relied on to dispose of CO2 in the United States.

Statoil has also captured CO2 from its production of natural gas at the Sleipner field off the coast of Norway. Starting in 1996, Statoil captured around 1 million tons of CO2 annually, and sequestered it in a nearby geologic formation.

As a result, Statoil has captured and sequestered a total of around 20 million tons of CO2.

The 20 million tons captured by Statoil over two decades is minuscule compared with the 2,298 million metric tons (MMT) produced every year from the generation of electricity in the United States.

Statoil has proposed making the sequestration of CO2 a business, but Europe, the main proponent of the CO2 hypothesis that CO2 causes global warming, hasn't adopted Statoil's proposal.

If Europe isn't adopting CCS, why should anyone believe it's viable?

There also has to be a guarantee that the millions and millions of tons of sequestered CO2 won't leak into the atmosphere over the next several thousand years.

Carbon Capture & Sequestration

No one has been able to establish that 2,298 million metric tons of CO_2 from the United States can be injected into geologic formations every year for hundreds of years, with any degree of certainty that it won't escape into the atmosphere. Let alone the multiple billions of tons of CO_2 from China, India and Europe.

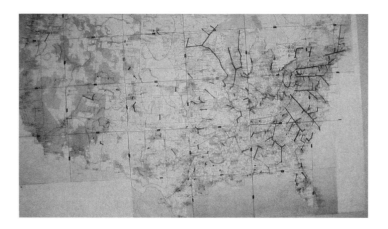

Figure 16. Map depicting probable location of CO_2 pipelines in the United States.

The locations of the pipelines in Figure 16 are based on the location of existing coal-fired power plants and does not include the additional pipelines required for natural gas power plants.

The pipelines are also based on the location of geologic formations depicted in the Carbon Sequestration Atlas of the United States and Canada, and assumes that these geologic formations are suitable for storing CO_2 for thousands of years.[33]

No one knows how to measure whether the sequestered CO_2 is leaking into the atmosphere or into adjoining geologic formations.

Carbon Capture & Sequestration

It's clear that CCS would be an undertaking of immense proportions, cost trillions of dollars without any assurance that the sequestered CO_2 would remain sealed underground and not leak into the atmosphere.

The United States could not act alone if CCS is to have any meaningful reduction of atmospheric CO_2.

China and India would also have to adopt CCS and sequester their CO_2 emissions underground.

If CO_2 emissions from power plants in China are the same percentage of total CO_2 emissions as in the United States, China would have to sequester around 8,000 million metric tons of CO_2 annually by 2030, adopting the MIT projection for China's CO_2 emissions in 2030.

And then there is Europe and the rest of the world.

A CO_2 pipeline map was drawn for Europe by CCS TlM, depicting the location of pipelines in Europe in 2050 to transport CO_2 to where it might be sequestered.[34] See Figure 17.

Who can believe that China and other developing nations are going to divert trillions of dollars from their economies to adopt CCS?

Those are dollars that could otherwise be used to achieve economic growth and rescue millions from poverty.

CCS is one of those fantasies that people like to use to weave beautiful scenarios in support of impossible projections.

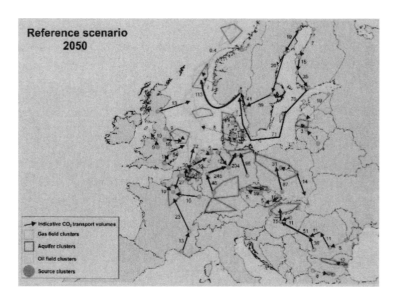

Figure 17. Map depicting CO2 pipelines in Europe in 2050 with a total length of approximately 13,000 miles.[25]

Chapter 13

Waxman-Markey

The American Clean Energy and Security Act of 2009 (ACES), known as the Waxman-Markey bill after its sponsors, passed the house, but was defeated in the Senate.[35]

While the Waxman-Markey bill did not become the law of the land and is old news, it's an example of how far supporters of the CO_2 hypothesis will go to control the daily lives of Americans.

Waxman-Markey is proof that supporters of the CO_2 hypothesis are willing to go to any length, including usurping the rights of all Americans, to enact laws and regulations controlling nearly every aspect of life.

It is, as some might say, the smoking gun.

The legislation was over 1,400 pages long, with numerous references to existing laws that would be modified by incorporating new language, or modifying language in existing legislation.

Reading the legislation was a nightmare, so it's doubtful very many members of Congress who voted for the legislation ever read it.

In fact, it was necessary for the House of Representatives to pass an amendment after passing the legislation to correct many errors in the original legislation as passed by the House.

The primary purpose of Waxman-Markey was to establish a cap & trade system for controlling CO_2 and other GHG gasses.

Waxman-Markey

Toward that end, Waxman-Markey established programs for cutting CO2 and other GHG emissions from vehicles, power plants and other sources, and gave the EPA authority to implement them.

Here's an example of how Waxman-Markey delved into detail that was incomprehensible.

Section 610 (a) (13) (B) iii concerned whether electric generation equipment could be installed on an existing dam.

The relevant text read:

> "The hydroelectric project installed on the dam is operated so that the water surface elevation at any given location and time that would have occurred in the absence of the hydroelectric project is maintained, ..."

A strict reading of this sentence would say that electricity could never be produced at such an installation.

When water flow increases, such as when it rains, water levels would rise, but if the hydro generators are put into operation, the water levels can't rise as much as they would otherwise have risen. Even if the difference can't be measured, which is possible on a large river, the theoretical result is obvious ... water levels would have risen by some additional fraction of an inch if the generators hadn't been placed into service.

The wording was probably intended to prevent water levels from being lowered, but even that would negate the ability to generate very much electricity as power generation would be limited to the amount of rainfall.

Waxman-Markey

Lawyers would have had a heyday with Waxman-Markey, and all Americans would have suffered.

But Waxman-Markey went far beyond this, and established legislation covering everything from lightbulbs to toilets. Regulations would apply to incandescent bulbs, fluorescent lamps, cool roofs, air-conditioning, proper ventilation (commercial buildings), electric motors, and for carbon content labeling, including for "iron, steel, aluminum, cement, chemicals, paper products, food, beverage, hygiene, cleaning, household cleaners, construction, metals, clothing, semiconductors and consumer electronics."

Here is a typical section of the Bill.

"ZERO-NET-ENERGY BUILDINGS. — In setting targets under this subsection, the Secretary shall consider ways to support the deployment of distributed renewable energy technology, and shall seek to achieve the goal of zero-net-energy commercial buildings established in section 422 of the Energy Independence and Security Act of 2007 (42 U.S.C. 17082)."

Here is an example of proposals contained in Waxman-Markey that would have infringed on the rights of Americans.

Waxman-Markey, HR 2454, required that new and substantially renovated commercial and residential buildings achieve a total reduction in energy use of 70% by 2030. (Substantially renovated would mean anything the federal government dictated.)

HR2454:

* Accomplished this objective of energy reduction by requiring states to adopt a national standard building code.

* Required the federal government to establish ways in which to measure whether the required reductions were achieved.

* Established that inspectors would tell local building contractors and architects how they should change their designs so as to meet the national energy use reduction target.

* Required state compliance plans to include the hiring of enforcement staff.

State governments would have been required to certify to the federal government that at least 80% of its urban population was covered by the national code. If states did not so certify, Waxman-Markey gave the federal government the right to impose the national standards without local agreement.

Waxman-Markey addressed residential issues by requiring new and existing homes to be audited to determine whether they met energy standards.

It established the Retrofit for Energy and Environmental Performance (REEP) program that addressed both residential and commercial buildings.

Section 202 included the following:

"The Administrator shall develop and implement, in consultation with the Secretary of Energy, standards for a national energy and environmental building retrofit policy for single-family and multifamily residences."

Waxman-Markey mandated that States had to adhere to REEP if they were to receive funds from the program. In essence, all homeowners would have had to comply with these codes, especially when making modifications, and have their homes evaluated for energy efficiency. One proposal required the results of these audits to be attached to the deed.

For a practical matter, homeowners would have had to make the necessary modifications to their home to comply with the energy standards before attempting to sell their home, since buyers would not want to become potentially liable for any deficiency.

This is the case with Radon today.

In addition the Bill:

- Established a home energy rating system (HERS).

- Established a model building energy label for display on buildings.

- Established a label for water efficient products.

- Established Green Building standards and "establish[ed] incentives to encourage each such organization to provide that any such structures and buildings comply with the energy efficiency and conservation standards, and the green building standards, under section 284(b) of such Act."

- Required the labeling of products to show their carbon content.

Waxman-Markey

Waxman-Markey would have led to a huge bureaucracy, first to define the regulations so they could be administered, and then to oversee them.

For example, Waxman-Markey required an initial 30% reduction in energy use by buildings.

Here's how the Pacific Northwest National Laboratory (PNNL) approached the problem of establishing base line energy usage around the country against which the 30% reduction could be determined.[36]

> "PNNL's methodology involves creating *16 prototype building designs* with which to model code factors. The Laboratory then simulates energy savings in these buildings in *17 climate zones* representing conditions throughout the nation. To find the national average energy savings, PNNL weights the individual results."

(Emphasis added.)

This proposed detail was for only one of the numerous mandates contained in the legislation.

Waxman-Markey delegated to the administrator, i.e., head of various agencies, the duty to establish and administer the detailed regulations required by Waxman-Markey.

The term "administrator" occurs 846 times in Waxman-Markey.

Here are a few things Waxman-Markey delegated to bureaucrats.

- The administrator shall promulgate regulations providing for the distribution of emission allowances allocated pursuant to section 782(f).

- The administrator shall establish regulations as to how allowances are to be distributed.

- The administrator shall establish design elements and requirements for electric vehicle programs.

- The Administrator shall distribute allowances allocated pursuant to section 782(g)(2) of the Clean Air Act to the SEED Account for each state.

The authority delegated to bureaucrats by Waxman-Markey was extensive, overreaching and appalling.

While Waxman-Markey was clumsy and heavy handed, a similar, far simpler scheme for controlling people's freedom was proposed for inclusion in a Climate Change Bill for the United Kingdom.

Here's how the scheme was described by the **Daily Mail**:

> One method could be personal carbon-allowances, where everyone is given a fixed amount of carbon to use each year. Each time they travel in a plane, buy petrol, go shopping or eat out would be recorded on a plastic card. The more frugal could sell spare carbon to those who want to indulge themselves. But if you were to run out of your carbon allowance, you could be barred from flying or driving.

Waxman-Markey

Waxman-Markey was too lengthy and cumbersome for a detailed analysis here, but these few examples clearly demonstrate that this type of legislation is an attack on personal liberty and freedom.

The EPA is, however, attempting to implement many of these same regulations without congressional approval, by claiming it already has approval under exiting law.

The entire purpose of this type of legislation is to restrict the use of energy, which is exactly opposite of what we should be doing.

Using energy can bring millions of people out of poverty, increase lifespans and generally improve the human condition.

Chapter 14

Impossible Objective

The UN, President Obama and the EPA have all said the United States must cut its CO2 emissions 80% by 2050, if the climate disaster is to be avoided.

Table 2, Chapter 5, reported U.S. CO2 emissions from each sector in 2004.

Some non-government organizations have championed major cuts in CO2 emissions. The Natural Resources Defense Council supports an 80% reduction in CO2 emissions from 1990 levels by 2050. Germany has targeted an 80% reduction from 1990 levels. The United Nations wants an 80% reduction for the U.S. so that developing countries, such as China and India, can continue to increase their CO2 emissions, but at a slower rate of increase. The Liberal Democrats of the U.K. have suggested it might be necessary to reduce CO2 emissions by an amount approaching 100% by 2050.

Clearly, when someone says CO2 emissions must be cut 100%, hysteria is running rampant.

Aside from whether there is a legitimate reason for cutting CO2 emissions 80%, there is the very real question of whether it is even possible to cut CO2 emissions 80%. There is also a question as to whether there would be adverse consequences if CO2 emissions were cut by that much.

The EPA, the Obama administration and the IPCC don't discus what it would mean for the United States to cut its emissions 80%.

Impossible Objective

Their commentaries are superficial and focus solely on the supposed need to avoid a climate catastrophe, but never on the consequences of cutting CO2 emissions.

What would it mean for Americans if CO2 emissions were cut 80% by 2050, only 35 years from now?

Table 4 shows how the U.S. cutting CO2 emissions 80% by 2050 would result in a dramatic cut in per person CO2 emissions from 16.6 metric tons to just 2.3.

Table 5		
Emissions, Population and Per Person Emissions 2012 — 2050		
	2012	**2050**
CO2 emissions in Metric Tons	5,200,000,000	1,080,000,000
U.S population	314,000,000	440,000,000
Metric Tons per person	**16.6**	**2.3**

In 2012, U.S. CO2 emissions were 5,200 million metric tons (MMT), while the population was 314 million. This equates to 16.6 tons per person.

Impossible Objective

The base year for all calculations and for all objectives established by the IPCC and the EPA etc. is the year 1990. The year 1990 continues to be used as the base year because all calculations relating to CO_2 reductions in the Kyoto treaty were based on reductions from 1990.

In 1990, U.S. CO_2 emissions were 5,040 MMT, while the population in 2050 is expected to be 440 million.

Cutting CO_2 emissions 80% by 3,960 MMT, brings total U.S. CO_2 emissions in 2050 to 1,080 MMT.

Taking the larger U.S. population into consideration, **CO_2 per capita emissions would be 2.3 tons**.

To put it in context, it's the level of U.S. per capita CO_2 emissions at the turn of the last century, i.e., 1900 … before the First World War.

There were very few automobiles, no commercial airplanes, no tractors and fossil-fueled power equipment for growing the food we need, no TVs, few electric lights, or any other powered equipment, such as refrigerators or air-conditioning units.

Here are some of the items invented or that were sparsely used until after 1900:

Automobiles
Airplanes
Air-conditioning
Aluminum foil, wiring
Arc welder
Blood tests
Buses

Impossible Objective

Cellophane
Cellular phones, iPhone
Compressed gasses, medical & industrial uses
Computers
Copiers
CT Scans
Dishwashers
Driverless cars
DVDs
Electric motors
Electric stoves
Electric washing machines and driers
Electric tools of all types, saws, screw drivers etc.
Electric fans
Electric vacuum cleaners
Elevators
Escalators
Fax
Flood lights for nighttime sports
Frozen foods
Gas turbines
Heating systems, forced hot air
Helicopter
Internet
Iron lung
Jet engines
Lazer
Lightbulbs
Microchips
Microwave ovens
Movies

MRIs
Nylon
Powered lawn mowers
Plastics, Bakelite
PV solar
Radios
Radar
Refrigerators
Robots
Sonar
Sonograms
Streetlights
Subways
Transistors
Trucks
TVs
Videos
Water purification with ozone
Water softeners using ion-exchangers
Xrays

This is only a partial list of all the inventions and products that were developed and widely used after 1900.

In other words, radical environmentalists, the IPCC, the Obama administration and the EPA want Americans to return to the "good ol" days of virtual poverty.

The approximate reductions achieved by segment when CO2 emissions are cut 80% are shown in Table 6.

An extreme environmentalist will look at this table and say, "No problem."

Table 6 80% Reduction in U.S. CO2 Emissions from 1990 levels by 2050 (in MMT).			
Source	**2004 Actual**	**% Total 2004 emissions**	**2050 Target 80% below 1990**
Electric Generation	2298.6	39%	421
Gasoline	1162.6	20%	216
Industrial	1069.3	18%	194
Transportation (Excluding Gasoline)	771.1	13%	140
Residential	374.7	6%	65
Commercial	228.8	4%	43
United States Total	5905.1	100%	1,080

Impossible Objective

He will say, wind and solar can replace all existing coal-fired power plants and nearly all the existing natural gas power plants. But as we have seen earlier, this is pure fantasy.

The answer from the extreme environmentalist will be the same for gasoline: "Just replace all gasoline vehicles with battery powered vehicles."

But this is absurd, since it will mean even more wind and solar to generate all the additional electricity required to recharge batteries, when wind and solar can't even replace all the coal-fired and natural gas power plants.

And the railroads? How will they cut CO_2 emissions 80%? Highway trucks use diesel fuel. How do they cut CO_2 emissions 80%? Horses can't haul big rigs.

And farm equipment? Do farmers go back to using horses? And construction equipment? Will handsaws have to replace power saws?

And what about airplanes? New jet engines may be able to cut emissions somewhat, but not by 80%.

And most of the residential emissions are from heating. Are we all supposed to live in cold homes since heating is done with natural gas or electricity? The same with hot water. Or do we cut down our forests to heat our homes?

And most commercial buildings use natural gas for heating and electricity for air-conditioning. How do they cut CO_2 emissions 80%?

The entire proposal to cut CO_2 emissions 80% is ludicrous.

Impossible Objective

And it will do no good to cut CO2 emissions by 30%, or any percentage below 80%, because the climate disaster can't be avoided if atmospheric CO2 levels go above 450 ppm, according to the IPCC's pseudo-science.

What's even more absurd is that even if the United States did cut its emissions 80%, China and the rest of the developing world will continue to spew CO2 into the atmosphere, making it impossible to keep atmospheric CO2 levels below 450 ppm.

And if the IPCC is right about CO2 causing global warming, which it isn't, it's impossible to avoid a climate disaster ... except by using energy.

Following the IPCC and president Obama's prescription would mean depriving the world of the energy it would need to respond to any changes in climate, either natural or man-made.

Chapter 15

An Alternative Hypothesis

There are several alternative hypotheses about global warming, but one that is gaining some prominence involves the sun and clouds.

The book, *The Neglected Sun* explores several alternatives. Quoting the authors[37]:

> There is no question that CO2, methane and other climate gasses have a limited warming effect on our climate. But there is also no doubt that a large part of the warming measured so far can be traced back to natural causes, with the sun having the most powerful impact on our climate.

Obviously, life on earth depends on the sun, but does the sun affect the earth in unique, perhaps not well understood, ways?

Light from the sun, i.e., photons, being converted to electricity, is understood and widely accepted as being beneficial.

The aurora borealis, a thing of beauty, is caused by the sun.

But the aurora borealis is also a manifestation of the sun's power.

The aurora borealis is caused by sun spots that produce huge changes in the magnetic fields around the sun and earth.

The largest known geomagnetic storm occurred in1859, and is known as the Carrington event. The 1859 storm was three times more intense than the most severe geomagnetic storm of the past thirty years.

An Alternative Hypothesis

The Carrington event took 17 hours, 40 minutes to reach the Earth, and it produced auroras seen around the world.

The Carrington event is vividly described in the book, *The Sun Kings,* by Stuart Clark. Highlights of how the Carrington event affected people around the world are mentioned here.

During the event, telegraph operators in Boston and Pittsburg found their equipment arcing, and were just barely able to disconnect the telegraph equipment from the lines.

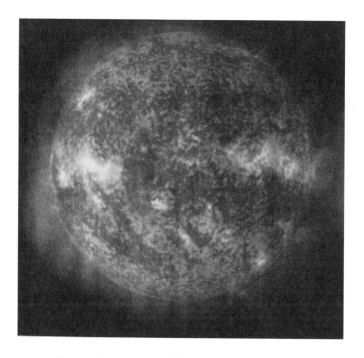

Figure 18. Photo of Sun Spots from NASA

Immediately after being disconnected the metal frames of the equipment were too hot to touch. The operator in Washington DC,

An Alternative Hypothesis

was stunned, and nearly killed, by an electric arc from the telegraph equipment that struck his forehead.

The current in the telegraph lines surged from nothing, to being so powerful that the telegraph keys were locked in a magnetic grip.

The aurora itself, in vivid displays of red and white, could be seen as far south as Key West, Florida, while intense streamers in the sky stretched from the South Pole far north into Chile.

The aurora came in two phases spread over two nights, and were so bright they could be seen in some areas during the daytime.

The Carrington event is significant because it occurred in 1859 when the only lines carrying electricity were telegraph lines.

Today, power lines are stretched across the United States, and across other countries, such as in Europe.

Could a sun spot 93 million miles from the earth affect the electrical grid, and all the people connected to it?

The answer became clear in 1989, when a geomagnetic storm caused the grid in Quebec, Canada to fail.

The 1989 storm was one-third the size of the Carrington event.

The reason for the grid failure in Canada was, quoting from the government's report:

> Ground induced currents (GICs) can overload the grid, causing severe voltage regulation problems and, potentially, widespread power outages. Moreover, GICs can cause intense internal heating in extra-high-voltage transformers, putting them at risk of failure or even permanent damage.

An Alternative Hypothesis

According to that same report, there are "300 EHV transformers in the United States" that are at risk.

A geomagnetic storm the size of a Carrington event could cause the grid to collapse if the EHV transformers fail, as they did in Canada, so that all the people in the northern part of the United States and in southern Canada would be without electricity for months, if not for over a year.

This demonstrates the raw power of the sun, and should give pause to those who say that the sun isn't affecting climate change.

It was Galileo, using his newly invented telescope, who, around 1600, saw sunspots for the first time in western history.

From that point forward, sunspot observations were made on a regular basis by astronomers throughout Europe.

Sunspot observations had also been made by the Chinese around 800 AD.

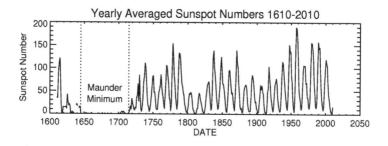

Figure 19. Sun Spot Cycles

In 1800, an astronomer, Herschel, was struck by the eleven-year sunspot cycle, and the perceived variation in commerce every

124

ten years. Turning to Adam Smith's *Wealth of Nations,* he found data about the price of wheat that varied with the sunspot cycle.

When sunspots were few in number, the price of wheat was high, and when sunspots were plentiful, there were abundant harvests and the price of wheat was low.

With this rudimentary idea that sunspots were related to climate, other astronomers, often sustaining criticism from their peers, searched for more data that could establish a stronger link between sunspots and climate.

Dramatic evidence of a strong linkage was provided by another astronomer, Walter Maunder, who at the age of 70, in 1922, linked the lack of sunspots between 1645 and 1715, to the bitter cold of that period.

It wasn't until the 1970s, when Dr. Jack Eddy focused attention on Maunder's work, that the significance of the Maunder Minimum became understood.

The Maunder Minimum is believed to have been the cause of the Little Ice Age. The Dalton Minimum is the period during the first two cycles beginning around 1800.

The 20th century seems to have been a period where sunspots were more frequent, especially from 1950 to 2000, while the most recent cycles in the 21st century have had fewer sunspots.

The forecast for cycle 25 is for fewer sun spots than in cycle 24, shown to the right of the NASA photo below. Cycle 24 has the fewest number of sunspots since cycle 14 that reached its peak around 1912.

An Alternative Hypothesis

Even if there is a correlation between sunspots and climate, it has only been recently that a mechanism, other than irradiance, has been proposed for the linkage.

In 1997, Dr. Svensmark, a Danish scientist at the Danish National Space Institute, proposed that sunspot eruptions affected the strength of the sun's magnetic field and the solar wind.

When the solar winds were strong, during periods of high sunspot activity, cosmic rays would be deflected away from the earth.

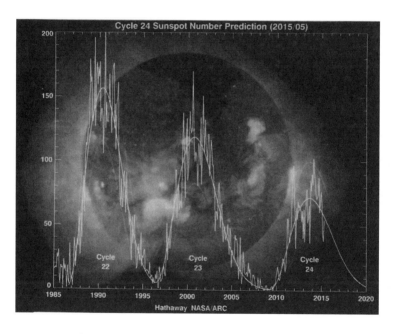

Figure 20. Sun Spot Cycles, from NOAA

When there were few sunspots, during periods of low sunspot activity, cosmic rays wouldn't be deflected away from the Earth

An Alternative Hypothesis

and could enter the earth's atmosphere, and affect the earth's climate.

Importantly, Svensmark suggested that cosmic rays could affect low level cloud formation.

He hypothesized that *more cosmic rays created more low level clouds*.

He proposed that an increase in low level cloud coverage would result in lower temperatures as they acted like a shade over the earth, while also reflecting more sunlight away from the earth's surface.

The major controversy surrounding Svensmark's hypothesis was whether cosmic rays could induce cloud formation.

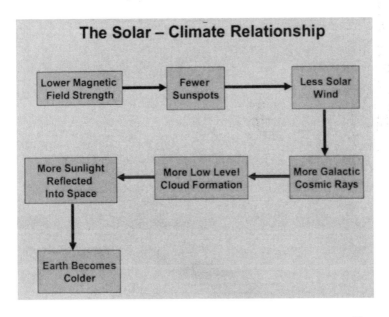

Diagram 5. Solar Relationships Schematic Diagram [38]

An Alternative Hypothesis

In 2007, Svensmark conducted a laboratory experiment that seemed to confirm galactic cosmic rays could induce cloud formation.

The debate then resulted in the Cloud experiment at CERN, Europe's premiere research center.

The Cloud experiment proved the amplification of aerosols. Additional observations, as described in the Neglected Sun, leave little doubt, that cosmic rays can induce cloud formation.[39]

Dr. Svensmark is upending the CO_2 hypothesis.

Dr. Svensmark has provided an explanation for how the sun, or more specifically sunspots, can affect climate change.

While this is admittedly only a hypothesis, it has substantial scientific underpinning going back several hundred years, and perhaps longer.

This is in contrast with the CO_2 hypothesis that's based on data going back only a hundred years or so.

In addition, the sunspot hypothesis is consistent with data for hundreds of years, at least back to 1600, while the data supporting the CO_2 hypothesis is not consistent.

From the mid-1800s throughout the 20th century, temperatures increased as atmospheric CO_2 increased, but prior to 1860, atmospheric CO_2 remained virtually constant while temperatures varied, up and down, including the Medieval Warm Period and the Little Ice Age.

If temperatures varied while atmospheric CO_2 remained constant, there cannot have been a very strong linkage between temperatures and atmospheric levels of CO_2.

An Alternative Hypothesis

The Carrington event demonstrated the power of the sun, while Svensmark has shown how its power could affect climate change.

Isn't the sun a more realistic answer for climate change than atmospheric CO_2?

Part 4

The Miracle of Fossil Fuels

"Using fossil fuels is not an addiction."

Alex Epstein
The Moral Case for Fossil Fuels

Chapter 16

Tragic War on Fossil Fuels

While it's becoming clearer every day that renewables can't replace fossil fuels, many elitists still condemn the use of fossil fuels. They mount campaigns to encourage pension funds and big investors to divest their ownership in fossil fuel companies. They campaign against fossil fuels in a war on affordable and reliable energy.

Figure 21. Fossil Fuel Protest Sign

But the truth is emerging, even among former supporters of wind and solar energy.

For example, two Google scientists, Ross Koningstein and David Fork, were put to work by Google in 2007 to establish that wind and solar energy could replace fossil fuels.

They were enthusiastic supporters of Google's project, known as RE<C, which was to develop renewable energy that would generate electricity more cheaply than coal-fired power plants.

But after only a few years, Koningstein and Fork admitted, in a November 2014 **IEEE Spectrum** article, what thousands of engineers and scientists had said all along, that it's impossible to cut CO_2 emissions using wind and solar to prevent a climate catastrophe, assuming CO_2 is the culprit.

Now, Bill Gates, a former huge financial backer of wind and solar projects, has changed his mind and admitted that wind and solar can't stop climate change.

In a 2015 **Financial Times** (FT) interview, Gates said, "Today's renewable-energy technologies aren't a viable solution for reducing CO_2 levels, and governments should divert their green subsidies into R&D aimed at better answers."

He went on to say, quoting from the FT, "The cost of using current renewables such as solar panels and wind farms to produce all or most power would be beyond astronomical."

While Gates intends to invest in R&D, he lambasted the prospects of battery technology, as it stands today.

He said, "There's no battery technology that's even close to allowing us to take all of our energy from renewables."

Tragic War on Fossil Fuels

We now have important former supporters of wind and solar admitting that wind and solar can't replace fossil fuels, and that subsidies for wind and solar should be stopped.

Yet, there is an unfolding tragedy around the world as fossil fuels are banned or economically black balled.

We need only turn to Africa to see how the war on fossil fuels fosters conditions that leave people living in poverty ... and dying at an early age.

The war on fossil fuels is literally killing people.

While natural gas is cleaner than coal, many parts of the world lack natural gas, but have large supplies of coal.

Sub-Saharan Africa is an area that is beset with poverty and short life expectancy.

Here is how McKinsey & Company describes the current situation.

> There is a direct correlation between economic growth and electricity supply. If sub-Saharan Africa is to fulfill its promise, it needs power — and lots of it. Sub-Saharan Africa is starved for electricity.
>
> It has 13 percent of the world's population, but 48 percent of the share of the global population without access to electricity.

While the average American consumes over 14,000 kWh/year of electricity, average consumption in the Central African Republic is 29 kWh/year, and in Chad it's only 8 kWh/year.

This is appalling, and tragic.

Tragic War on Fossil Fuels

For example, each person in Chad has enough electricity to be able to use two, 100 watt light bulbs for 10 hours a day. And, most people in Chad can't use any lightbulbs because electricity isn't available at their homes.

Energy access is defined by the International Energy Agency (IEA) as 250 kWh/year and 500 kWh/year, for rural and urban areas respectively.

Three resources are available in Sub-Saharan Africa that could be used to generate electricity — coal, natural gas and oil. All are fossil fuels.

Distribution of these resources is spotty. Natural gas is primarily in Nigeria, with some scattered in a few other countries, mainly the Congo, Namibia and Rwanda.

There are large reserves of coal in South Africa, with small reserves sprinkled throughout many other areas.

This coal could be used to generate electricity, but environmentalists are depriving African countries the money they need to build coal-fired power plants.

The Obama administration has announced it's cutting off funding for coal-fired plants overseas.

And, in a tragic change in its policy, the World Bank has also cut off funding for coal-fired plants around the world.[40]

It is a tragedy because it will condemn millions to live in poverty and die at an early age. In Sub-Saharan Africa alone, there are 600 million people without access to electricity.

Tragic War on Fossil Fuels

Pictures taken from outer space, such as the one shown in Figure 22, of Africa at night show a continent in dire need of electricity. Electricity that could be generated using fossil fuels.

While there are other problems in Africa, such as graft and religious warfare, the cheapest and most easily used resource is now unavailable for building power plants in Sub-Saharan Africa.

Power plants need water for cooling, and there are numerous rivers available for that purpose, with 90% of the cooling water withdrawn from rivers returned to the rivers.

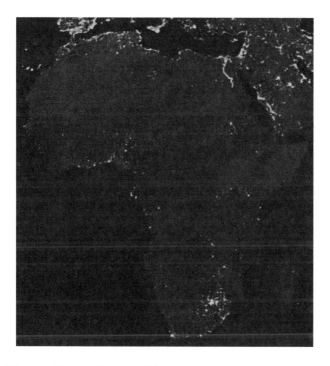

Figure 22. NASA Satellite Image of Africa at Night

Tragic War on Fossil Fuels

Wind farms are being built in Kenya, but they are a futile public relations effort by elites to show wind can provide electricity.

For example, the highly touted 300 MW Lake Turkana wind farm in Kenya cost $800 million, but it can only produce the same amount of electricity as a 100 MW coal-fired power plant.

More importantly, a natural gas power plant that can produce the same amount of electricity, reliably and without interruption, would have cost approximately $100 million. A similar coal-fired power plant would have cost approximately $250 million.

Not only is the Lake Turkana wind farm unable to generate very much electricity reliably, it wasted hundreds of millions of hard to-come-by dollars. These wasted dollars are dollars that poor countries desperately need.

Renewables, except for hydro in a few locations, such as the Congo, are unrealistic.

Nearly all governments are short of money, which is largely unavailable for very expensive renewable alternatives, such as the Lake Turkana wind farm.

When there are no gas turbines for back-up, wind is even more unreliable than it usually is.

Solar only generates electricity during the day, so African countries would go dark at night.

Without electricity, many, many African families are left cooking with dung and allowing wood and dung smoke to permeate their homes.

Tragic War on Fossil Fuels

While the situation in Africa is hard to comprehend, other countries are also targeted by the war on fossil fuels.

India and Indonesia, two very important and populous countries, need electricity.

In India, the average person consumes only 600 kWh/year, while in Indonesia, it's only 629 kWh/year.

Indonesia has large coal reserves for coal-fired power plants, and for export to sustain its economy.

Figure 23. Indonesia 2014, photo by D. Dears

India has large coal reserves which it is trying to develop, primarily so it can generate more electricity.

Do we expect these countries, with large populations, to relegate their citizens to continued poverty because of the war on fossil fuels?

Tragic War on Fossil Fuels

Similarly, China is developing its coal reserves and building new ultra-supercritical, highly efficient, power plants.

Without the participation of India, Indonesia and China, it's impossible to cut CO_2 emissions worldwide and prevent atmospheric CO_2 emissions from increasing.

The lack of electricity affects people in other ways.

Millions don't have access to drinking water.

While Africa, Asia and South America have many rivers that can provide cooling water for power plants, many locations lack the ability to distribute the water to communities located some distance from rivers.

Distributing water requires the use of pumps, driven either with electricity or fossil fuels, such as diesel or gasoline powered pumps and generators.

Irrigation of rice paddies and farms require water distributed to them by electricity or fossil fuel driven pumps. There are foot driven pumps, using a man walking on them, but that in itself is a tragic waste.

Consumers in countries lacking adequate supplies of electricity are rioting, upsetting the social order.

Everywhere, there are articles on how shortages of electricity are hampering growth and job creation.

The war on fossil fuels is a tragedy, because it's condemning people to poverty and an early death.

The use of fossil fuels is beneficial to mankind, and the war on fossil fuels is immoral.

Tragic War on Fossil Fuels

Mankind isn't addicted to fossil fuels. Instead, mankind needs fossil fuels to emerge from poverty and then sustain a healthy lifestyle.

Chapter 17

Advantages of Fossil Fuels

Fossil fuels are pilloried by environmental groups and many governments, especially the current administration of the United Sates and its EPA.

Diatribes against fossil fuels ignore all that humanity has gained from their use.

Before the use of fossil fuels "the life of man, was poor, nasty, brutish, and short.[41]"

It's only proper to take a minute to examine the benefits we have received form oil, coal and natural gas.

When the Sierra Club declared war on natural gas, they might just as well declared war on humanity.

Here are some of the ways that fossil fuels contribute to society.

Airplanes could not exist without jet fuel and gasoline derived from oil. There would be no travel by air, and companies such as Boeing would not exist.

Railroads could not exist without diesel fuel, coal or natural gas (LNG). The entire industry, including railroad companies, such as Union Pacific, and locomotive and car manufacturers, such as General Electric and Trinity Industries would not exist, including all the associated jobs.

Steel could not be made without coal. Without steel, sky scrapers and suspension bridges could not be built. Jet engines and gas turbines couldn't be built.

Advantages of Fossil Fuels

Roads capable of withstanding the heavy traffic of speeding cars and heavy trucks couldn't be built without asphalt or cement. Asphalt is derived from oil. Producing cement requires the use of oil, natural gas or coal.

Cobble stones might be able to stand the wear and tear of traffic, but would cause incredible damage to vehicles and their cargo from the jarring they would receive.

Many plastics and other chemicals couldn't be produced without oil, natural gas or coal. Some plastics, such as PVC pipe, and some textiles, such as nylon, and some solvents, cosmetics and pharmaceuticals couldn't be made without oil, natural gas or coal. CDs, DVDs and vinyl records are made from oil.

Refrigerators and air conditioning units require steel or aluminum and a chemical refrigerant. Oil, natural gas or coal are required at some point in the manufacture of these products.

Glass for windows, bottles, cars and everyday glassware require the use of oil, natural gas or coal.

Heating of homes and buildings require oil, natural gas or coal.

Other products that require oil, natural gas or coal at some point during their manufacture include cans, for canned food, the manufacture of copper wire, essential for the electrical industry, the manufacture of automobiles, trucks, earthmoving equipment and even toilets.

For a few of these items, it might be possible to substitute wood, but it's doubtful there would be enough trees to meet all these requirements. Mountainsides are being denuded in Africa by people using wood to heat their homes and cook their food.

Advantages of Fossil Fuels

Forests were being denuded in Europe before the discovery of fossil fuels.

Mining companies that mine coal, drilling companies that produce oil and natural gas are providing a vital service to society.

Yet some people and environmental organizations want to have these companies divested from pension funds.

Two examples of how fossil fuels have helped humanity may be of interest.

At the turn of the last century, most transportation used horses.

The residents of New York City, including Brooklyn, used around 200,000 horses and they produced four million pounds of manure every day, plus millions of gallons of urine.

At the time, each horse consumed the product of five acres of land, a footprint which could have produced enough food to feed six to eight people.

Sweepers were stationed at street corners to sweep a path through the manure so people could cross the streets.

Manure and urine turned to muck when it rained, but conditions were actually worse during dry weather when the muck dried out and turned to dust that was whipped up and down the streets by the wind, chocking pedestrians and coating buildings.

Thousands died each year in New York City from diseases carried by flies, but it's difficult to know exactly how many because so many were dying from heat in the summer and from food spoilage due to inadequate cooling from ice boxes.

Every day New York City had to get rid of 40 or more dead horses that had been abandoned by their owners. Some of these were allowed to decay as it was easier to cut up and move a partially decayed horse.

The innovation that curtailed the use of horses for transportation was the automobile powered by fossil fuels.

Another innovation involving fossil fuels occurred in the mid 1800s.

During the eighteen and nineteenth centuries, oil from sperm whales became the preferred method for lighting homes, with richer people also preferring candles made from spermaceti found in the sperm whale's nose.

Sperm whales were hunted mercilessly, with each large whale producing as much as 3 tons of oil.

It's estimated that more than 200,000 sperm whales were killed during the 1800s.

If this had continued, there's little doubt that sperm whales would have become extinct.

The invention of Kerosene produced from oil in 1849 upended the whaling industry.

The price of sperm oil fell from a high of $1.77 per gallon in 1856, to around 40 cents.

Meanwhile, the price of kerosene fell to 7 cents per gallon by 1895.

Kerosene, a fossil fuel, had saved the sperm whale.

The list of benefits derived from fossil fuels is endless, including low-cost electricity that can't be reliably produced in large quantities by wind or solar.

Advantages of Fossil Fuels

The book, **The Moral Case For Fossil Fuels**, by Alex Epstein, provides a far more complete argument for why fossil fuels are beneficial to mankind.

In short, fossil fuels are essential to modern day living, maintaining our standard of living and freeing millions from poverty.

The war on fossil fuels is a war on humanity.

It is simply another reason why attempts to cut CO_2 emissions is a fool's errand.

Thought should be given to the benefits of fossil fuels before we pillory them and the companies that produce them.

Chapter 18

Remarkable Availability of Life-Saving Fossil Fuels

Mankind is blessed to have an abundant, readily available and inexpensive supply of fossil fuels: A supply that may last for a thousand or more years.

Coal, oil and natural gas are widely available around the world. None is in danger of being depleted.

Any misconceptions based on "peak oil" have been soundly discredited with the development of fracking.

Coal is probably the easiest of the three to extract and use. It is also widely available.

It has served mankind well, especially as new technologies have been developed for burning coal safely and for capturing any pollutants that are emitted in the process.

As each nation develops its economy, it puts these new coal burning technologies into service.

Today, ultra-supercritical coal-fired power plants are 45% more efficient than the units in use over the past half century. These new, more efficient units are being built in China and India, but can't be built in the United States because of EPA regulations limiting CO_2 emissions. See Appendix 2.

While established worldwide coal reserves can last for over 100 years, there is undoubtedly far more coal to be found. At present, few people are exploring for coal, but as has been proven true for all other reserves, new reserves are bound to be found so that coal can probably last for centuries.

The same is true for oil. While proven reserves can supply the world with oil for over 50 years vast areas of the globe have yet to be explored.[42] These conventional reserves are only part of the available oil that can be found in shale around the globe.

Worldwide shale oil reserves may be 5 trillion barrels, while proven conventional oil reserves are 1.3 trillion barrels. While only some of the 5 trillion barrels of shale oil may be recoverable, it's clear that oil will be available for many years beyond this century.

The truly remarkable fossil fuel resource is natural gas.

According to the Potential Gas Committee, "The United States possesses a total technically recoverable resource base of 2,515 trillion cubic feet (Tcf) of natural gas as of year-end 2014."

This is the highest resource evaluation by the Potential Gas Committee in its 50-year history.

The reserves would last over 100 years if new uses weren't added to current uses.

The impact of major new uses could affect the supply of natural gas in the United States in the following ways.

- Converting 50% of its coal-fired power plants to natural gas will use an additional 5.3 Tcf per year.

- Exporting LNG from all proposed 19 terminals will use an additional 10.4 Tcf each year.

- Converting 100% of long haul trucks to natural gas will use an additional 7.3 Tcf per year.

Remarkable Availability of Life-Saving Fossil Fuels

Including these additional uses results in the U.S. having a 50+ year supply of natural gas.

But this is a static picture of currently recoverable natural gas reserves.

It's merely a snapshot in time.

Historically, recoverable reserves have increased yearly as the result of new technologies being used to extract oil and gas from the ground.

There is no reason to believe this process of continually improving technologies will end in the future.

In fact, new fracking techniques are already increasing how much oil and natural gas can be extracted from shale formations.

Longer, more closely spaced laterals, increased number of fracs and increased quantities of frac sand has resulted in increased output from each well, and from each shale formation.

This will mean that recoverable reserves will increase each year into the future, for an unknown number of years.

It could easily mean we have natural gas reserves that will last beyond this century, probably for an additional one or two hundred years, and possibly for a 1,000 years.

Unquestionably, shale gas will be extracted from shale formations around the world, including Argentina, the UK and China.

Remarkable Availability of Life-Saving Fossil Fuels

The age of coal will not immediately go away. Huge worldwide reserves of cheap coal remain in many countries, including Indonesia, India and China. But, as this projection from the EIA shows, Figure 24, natural gas will probably surpass the use of coal by the end of this century.

While natural gas from conventional and unconventional sources may last a thousand years, methane hydrates could provide natural gas worldwide for an additional thousand years.

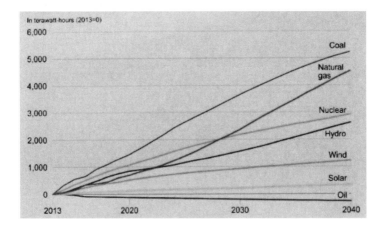

Figure 24. Fuel Usage From Energy Information Administration

Table 7 depicts available methane hydrates on the U.S. outer continental shelf (OCS), which is only a small fraction of methane hydrates available worldwide.

The total amount of methane hydrates along the outer continental shelf of the United States is twenty times current U.S. natural gas reserves.

Remarkable Availability of Life-Saving Fossil Fuels

Both Japan and the United States have already begun developing methods for extracting natural gas from methane hydrates.

Table 7	
Region	**In-place Gas Hydrate Resources**
Atlantic OCS	21,702 Tcf
Gulf of Mexico OCS	21,444 Tcf
West Coast OCS	8,192 Tcf
Total	51,338 Tcf

An attractive aspect of natural gas is that it can be converted into other chemical forms.

Natural gas is flexible.

Not only can it be converted into other liquid fuels, such as diesel fuel, it can be liquified and compressed so as to be readily transportable for use in trucks, river transportation and ships.

In addition, natural gas is the feed stock for chemical compounds.

It can be used to make plastics, yarns for fabrics and also fertilizer.

It is even used in the manufacture of pharmaceuticals.

Natural gas is truly the wonder resource.

Just think: Natural gas can heat our homes, power our cars and trucks, and help grow the food a hungry world needs.

And mankind will have it available for a long, long time.

Conclusion

Americans deserve the truth about CO2, global warming and climate change.

Everything in this book is factually correct.

So why is it that much of the information in this book is contrary to what is routinely reported in the media?

Why hasn't the media investigated the misinformation published by the EPA, NASA, NOAA and other government agencies?

Those proclaiming that CO2 emissions are the cause of global warming and climate change should be required to debate those who believe the science does not support such claims.

Instead, the supporters of climate change alarmism hide behind an assertion that the science is settled and that there is no need to debate with Flat Earther's, or some similar snub.

Yet, it is Americans who suffer as the result of this type of brush off.

It is Americans who see their electric bills increasing every year because of the policies put in place to cut CO2 emissions.

It is Americans who have their tax money used to subsidize renewables and programs whose only purpose is to cut CO2 emissions.

It is Americans who are losing their jobs because fossil fuels are being systematically outlawed.

Today it's coal miners and those working in related industries who are losing their jobs. Tomorrow it will be those who work in

Conclusion

the oil and natural gas industries who will lose their jobs because these fossil fuels also emit CO_2.

While this text merely scratches the surface of the science of global warming, it uses engineering principles and logic to establish the flaws in the CO_2 global warming hypothesis and related proposals for cutting CO_2 emissions.

Over 31,000 engineers and scientists have signed the petition that states there is no scientific evidence to support the claim that CO_2 emissions could cause catastrophic climate change.[43]

A group of scientists, the Nongovernmental International Panel on Climate Change (NIPCC), has published a series of scholarly reports that provide exhaustive information on the science of climate change and why the alarmists views are in error. The reports are available for any reader who wants to delve into the scientific details of the subject.[44]

Americans deserve the truth about CO_2, global warming and climate change, and it's hoped this book will encourage them to not only seek it, but to insist that the media and government agencies provide it.

Appendix 1

Time required to recover an investment in a PV rooftop solar installation

The following tables show the number of years it would take to for a homeowner to recover an investment in a PV rooftop solar installation in each state, *without tax credits*.

The calculations were performed using an Internet calculator provided by Anapode Solar during July 2015. Installation labor is not included in the calculations. The added cost of installing the system could increase the number of years required to recover the cost of the system.

Electricity costs are the average for each state. Residential rates by individual utilities will likely be different.

The calculation assumed electricity usage of 1,500 kWh per month.

The paybacks assume the PV rooftop installation faces due south. If it faces east or west, it will increase the payback time by an additional 2 or more years.

Except for Hawaii, none of the installations produce a good rate of return.

Virtually all installations require ten or more years to recover the investment. Recovery periods are significantly affected by the electricity rates in each state.

In a few instances, denoted by an asterisk, the calculator did not show a particular state. The nearest location was used for the calculation.

Appendix 1

The cost of PV rooftop solar systems is likely to come down somewhat over the next decade, but even if costs are cut by 50%, there would be very few locations in the United State where PV rooftop solar would be a good economic investment.

Appendix 1

Years to recover cost by state					
State	Years	Cost of Installation (without tax credits)	Annual Savings	% of Electric Bill Covered	Cents /kWh
Alabama	18	$30,731	$1,748	104%	9.30
Alaska	10	$30,731	$3,110	98%	17.58
Arizona	10	$16,400	$1,593	98%	10.24
Arkansas	23	$30,731	$1,327	94%	7.85
California, LA	8	$20,733	$2,533	92%	15.23
Colorado	14	$25,720	$1,809	100%	10.04
Connecticut*	11	$30,731	$2,891	95%	16.98
Delaware*	15	$30,731	$2,098	104%	11.33
Florida	11	$20,733	$1,819	93%	10.86
Georgia	15	$25,719	$1,737	97%	9.94
Hawaii	3	$20,733	$5,959	99%	33.53
Idaho	18	$25,719	$1,452	101%	7.95
Illinois	25	$30,731	$1,233	77%	8.87
Indiana	18	$30,731	$1,678	104%	8.97
Iowa	19	$30,731	$1,611	108%	8.24
Kansas	12	$20,733	$1,721	95%	10.04
Kentucky	17	$25,719	$1,483	101%	8.13
Louisiana	19	$25,719	$1,390	95%	8.11
Maine	12	$25,719	$2,108	93%	12.66
Maryland	13	$25,719	$2,001	92%	12.12

Appendix 1

Years to recover cost by state					
State	Years	Cost of Installation (without tax credits)	Annual Savings	% of Electric Bill Covered	Cents /kWh
Massachusetts	12	$30,731	$2,607	95%	15.34
Michigan	16	$30,731	$1,960	98%	11.10
Minnesota	16	$25,719	$1,605	93%	9.63
Mississippi	16	$30,731	$1,896	109%	9.66
Missouri	17	$30,731	$1,758	108%	9.06
Montana	16	$25,719	$1,564	101%	8.62
Nebraska	17	$25,719	$1,555	98%	8.80
Nevada	11	$20,732	$1,842	105%	9.76
New Hampshire*	12	$30,731	$2,596	95%	15.25
New Jersey	12	$30,731	$2,610	104%	14.01
New Mexico	11	$18,700	$1,739	100%	9.69
New York	12	$30,731	$2,557	87%	16.25
North Carolina	16	$25,719	$1,622	97%	9.32
North Dakota	16	$25,719	$1,565	103%	8.49
Ohio	17	$30,731	$1,838	102%	9.97
Oklahoma	15	$20,733	$1,342	92%	8.10
Oregon	18	$25,719	$1,464	93%	8.78
Pennsylvania	21	$30,731	$1,491	81%	10.29
Rhode Island	11	$30,731	$2,922	104%	15.57
South Carolina	14	$25,719	$1,780	104%	9.56

Appendix 1

Years to recover cost by state					
State	Years	Cost of Installation (without tax credits)	Annual Savings	% of Electric Bill Covered	Cents /kWh
South Dakota	15	$25,719	$1,724	107%	9.06
Tennessee	16	$25,719	$1,565	91%	9.50
Texas	15	$25,719	$1,760	109%	8.99
Utah	16	$25,719	$1,630	108%	8.41
Vermont*	13	$30,731	$2,295	87%	14.58
Virginia	18	$30,731	$1,695	102%	9.25
Washington, Spokane	22	$25,719	$1,179	92%	7.15
West Virginia	25	$30,731	$1,239	90%	7.65
Wisconsin	15	$30,731	$2,039	106%	10.73
Wyoming	15	$20,733	$1,395	99%	7.78

Appendix 2

Ultra-supercritical coal-fired power plants

The existing fleet of coal-fired power plants in the United States, and most elsewhere around the world, has a thermal efficiency of 32%, using higher heating value (HHV).[45]

New ultra-supercritical pulverized coal-fired power plants are an important improvement over traditional pulverized coal-fired plants.

Ultra-supercritical plants have a thermal efficiency of 44% HHV, which is a 38% improvement over traditional plants.

Ultra-supercritical plants with an efficiency of 44% operate at very high temperatures and pressures, typically 1112° F and 4350 psi.

It's anticipated that temperatures and pressures can be increased further, and that a thermal efficiency of 46% HHV can be achieved in the next several years. These units would be referred to as Advanced Ultra-supercritical plants.

Both the United States and Europe have development programs designed to produce the metals needed for boilers and steam turbines to operate at temperatures and pressures higher than these.

Ultra-supercritical coal-fired power plants, with appropriate controls for capturing particulates and mercury, can operate with extremely low levels of NOx, SOx.

China is building a large number of these ultra-supercritical coal-fired power plants.

Appendix 2

The United States has built one ultra-supercritical plant, the Turk facility in Arkansas, but is prohibited by EPA regulations limiting CO_2 emissions from building any more such units.

Appendix 3

The decline of nuclear power in the United States

The United States currently has 100 nuclear power plants in operation. Four new nuclear plants are under construction: Two in Georgia and two in South Carolina.

These 104 plants will continue to produce about 20% of U.S. electricity ... for a few more years, and then the decline begins.

Existing nuclear power plants need to receive a 20-year renewal to their original 40-year operating license.

Approximately 87 of the 100 existing plants have received their 20-year license renewals, and it has been widely assumed the remaining units will also receive renewals, though a few are now in question due to environmental agitation.

Importantly, all existing nuclear power plants will have to obtain a second 20-year renewal when the initial 20-year renewal expires.

The first of the units that obtained their initial 20 year renewal will need to obtain their second 20-year renewal in the mid-2030s, about twenty years from now.

While obtaining the first 20-year renewal was reasonable, a second 20-year renewal may be problematic. At the end of a second 20-year renewal these nuclear power plants will be 80 years old, and it's logical to believe that these plants will be wearing out.

Nothing lasts forever, and everything from embrittlement of the reactor containment vessel to aging piping, valves and control systems could be cause for concern.

There is every reason to believe that most of these plants will not receive a second 20-year license renewal.

Without a second 20-year renewal, existing nuclear power plants will have to begin shutting down in the mid-2030s.

Unless new nuclear power plants are built, the amount of electricity supplied from nuclear power plants in the United States will begin to rapidly decline.

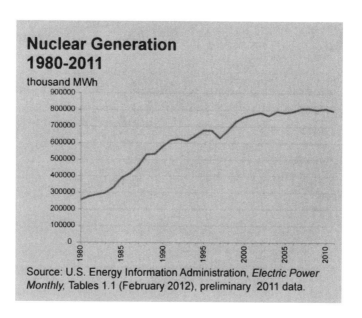

Nuclear Generation 1980-2011

thousand MWh

Source: U.S. Energy Information Administration, *Electric Power Monthly*, Tables 1.1 (February 2012), preliminary 2011 data.

The cost of new nuclear plants has grown to a staggering $6,000 / KW.

With low natural gas prices, and the possibility that new ultra-supercritical coal-fired power plants can be built at less than half

Appendix 3

the cost of nuclear power plants, the economics mitigate against building new nuclear plants in the United States.

With Yucca Mountain storage seemingly going nowhere, and with the public emotionally opposed to nuclear power, it would seem that nuclear will be in terminal decline in the United States.

There's a very strong possibility that nuclear will supply less than 5% of our electricity by 2100.

Even small modular nuclear power plants cost $6,000 / KW, so they may also be too expensive when compared with alternatives.

Meanwhile, the situation elsewhere in the world is different.

A total of 64,000 MW of nuclear power plants is being built elsewhere in the world, which translates into approximately 64 new nuclear power plants if their average size is 1,000 MW.

China is the largest builder of new nuclear power plants, with approximately 28 under construction.

Russia has about 4 plants under construction.

Interestingly, Russia's Rosatom is building 15 plants in other countries including Turkey, Vietnam, Belarus, and a few others.

India is building 4 or 5 units.

S. Korea is building 5 units.

It's highly probable that nuclear power in The United States will begin its long-term decline in around 20 years.

Notes

1. IPCC web site: http://www.ipcc.ch/organization/organization_history.shtml

2. ibid

3. Chapter 4, The Global Warming Scam and the Climate Change Superscam, by Vincent Gray, Ph.D. http://bit.ly/1NryeWf

4. Essay by Dr. Tim Ball, 2013, *Why and How the IPCC Demonized CO2 with Manufactured Information.*

5. *ibid*

6. AR5 Summary for Policy Makers and [U.S.] Presidential Action Project to stabilize climate.

7. Watts Up With That http://bit.ly/1AyH1yo

8. Watts Up With That http://bit.ly/1FSLqhA

9. L.B. Hagen, Roanokeslant

10. Watts Up With That, Paleoclimate Page http://bit.ly/1Dhip0O

11. Validation Of A Climate Model Is Mandatory: The Invaluable work of Dr. Vincent Gray, Ph.D., by Dr. Tim Ball, August 22, 2015 http://bit.ly/1OR9VzI

12. Charting the outcome of the Obama – China climate deal by 2030, by Hoskins http://bit.ly/1UO4wey

13. The EIA has calculated the LCOE for land based wind at 8 cents per kWh. This is based on some false assumptions that, if corrected, would result in an LCOE of over 10 cents

per kWh. For example, a 35% capacity factor is used, but the actual capacity factor of existing wind turbines is 32% or less. In addition nothing is included for the cost of backup power for when the wind stops blowing.

14. Congressional Research Service, *The Renewable Electricity Production Tax Credit: In Brief, 2014*

15. New York Times, New Rules and Old Plants May Strain Summer Energy Supplies, Aug 11, 2011

16. Eastern Wind Integration and Transmission Study by EnerNex Corporation, January 2010

17. LCOE from EIA, Annual Energy Outlook 2015

18. University of California, Sand Diego, Do The Math, Don't be an Efficiency Snob, http://bit.ly/1F6RE17

19. Lawmakers in many states require the utility to pay the PV rooftop owner the same rate for the electricity generated by the rooftop installation and sent to the grid, as the owner pays the utility for electricity taken from the grid. Typically this is 12 cents per kWh. Occasionally, the net metering payment is limited to the cost the utility would have incurred in generating the electricity at its power plants. Typically this is 5 cents per kWh. The difference of 7 cents per kWh represents overhead, including construction and maintenance of transmission and distribution lines, and profit, all of which is lost to the utility.

20. Roger Andrews analysis is at http://bit.ly/1T4UGqK

Notes

21. FERC Clarifies and Reaffirms Order on Bonneville Wind Curtailment "Oversupply Management" Policy, http://bit.ly/18oCjK6

22. Renewable Electricity Generation in Germany, from volker-quaschning.de http://bit.ly/1ieRhHj
The use of biomass was omitted form percentage calculation as it is merely a substitute for coal or natural gas in thermal power plants.

23. Energy Darwinism, The Evolution of the Energy Industry, CiTi GPS, October 2013

24. From EIA LCOE calculations web site http://1.usa.gov/1M8odN1

25. Elon Musk introduces Powerwall Battery, http://www.teslamotors.com/powerwall

26. EIA http://1.usa.gov/1WabsoS

27. From chapter 10. Clemson Extension, March 2014. Timberland Value: From Inventory Value to Cash Flows, by Thomas J. Straker

28. With an annual output of 16 million gallons of biofuel requiring 175,000 tons of pulp, and with annual jet fuel usage of 116 billion gallons: Divide 116 billion gallons by 16 million gallons and multiply by 175,000 tons of pulp = 1.3 billion tons of pulp.

    To determine the number of acres: divide 1.3 billion tons by 140 tons per acre = 9.1 million acres.

29. U.S. News & World Report, http://bit.ly/1Mymxyr

30. It's not possible to accurately determine the amount of electricity generated by PV rooftop solar because the number of hours of usage, weather and variations in insolation are not known. Comparing the SEIA estimate of installed capacity of 6,200 MW (2014) with EIA's estimate of installed capacity of utility scale solar of 6,674 MW (2013), it's reasonable to conclude that consumption from PV rooftop solar installations is approximately the same as from utility solar installations, which was approximately 0.4 quadrillion BTUs in 2014. See EIA http://1.usa.gov/1QqfVjx and NREL http://1.usa.gov/1NKMWta

31. Referring to Table 3, total energy produced by fossil fuels equals 80.4 quadrillion BTUs, and total energy produced by renewables equals 6.1 quadrillion BTUs. Referring to Figure 15, fossil fuels received approximately $3 billion in subsidies, while renewables received approximately $10 billion. Divide $3 billion by 80.4, and $10 billion by 6.1 to give dollars per quadrillion BTU, then divide the answer for renewables by the answer for fossil fuels to arrive at 44 times more subsidies for renewables than for fossil fuels.

32. PNNL.gov 2009 Future CO_2 Pipeline Not as Onerous as Some Think http://1.usa.gov/1IQz11h

33. Carbon Sequestration Atlas of the United State sand Canada http://bit.ly/1AcB32w p

34. Finally, the plan for the CCS revolution, by CCS TLM http://bit.ly/1wlAL9s

35. HR2454 http://1.usa.gov/1FIL24u

36. PNNL link http://1.usa.gov/1QKwRpo

37. *The Neglected Sun, Why the Sun Precludes Climate Catastrophe,* by Vahrenholt and Luning, 2015

38. From Watts Up With That http://bit.ly/1He8aKq

39. ibid

40. Reuters, July 16, 2013

41. Quotation from Thomas Hobbes' Leviathan, 1651

42. Total world proved oil reserves end of 2014, can meet 52.5 years of global production, from BP.com

43. Global Warming Petition Project, 31,487 American scientists have signed this petition, including 9,029 with PhDs, http://www.petitionproject.org

44. Nongovernmental International Panel on Climate Change, http://climatechangereconsidered.org

45. Appendix, National Coal Council Issue Paper 2009, Higher Efficiency Power Generation Reduces Emissions.

About The Author

Donn Dears began his career at General Electric testing large steam turbines and generators used by utilities to generate electricity; followed next, by manufacturing and marketing assignments at the Transformer Division. He led an organization of a few thousand people servicing these and other GE products in the United States. He then established facilities around the world to service power generation, transmission equipment and other electric apparatus. Later, he led an engineering department of several hundred people that provided engineering support to nearly a hundred service installations around the world.

At nearly every step, Donn was involved with the work done at customer locations: at steel mills, electric utilities, refineries, oil drilling and production facilities and open pit and underground mining operations. At every opportunity, he learned of the needs of these industries.

Donn has had a close-up view of the eastern province of Saudi Arabia with its oil producing and shipping facilities. He has investigated many of the other oil producing countries in the Mideast and Northern Europe, as well as examining iron-ore mining locations and major shipping centers in Europe and Asia. All told, Donn has visited over 50 countries and has knowledge of their need for the technologies that can improve their well being and their use of equipment manufactured in the United States.

Following his retirement as a senior GE Company executive, he continued to study and write about energy issues.

Donn is a graduate of the United States Merchant Marine Academy and served on active duty in the U.S. Navy.

Index

Index

Index

Index

Index